BUILDING WITH PAPER

BIRKHÄUSER
BASEL

BUILDING WITH PAPER

ARCHITECTURE AND CONSTRUCTION

ULRICH KNAACK REBECCA BACH SAMUEL SCHABEL (EDS.)

The editors would like to thank the research funding programme LOEWE of the State of Hesse and the Technical University of Darmstadt for their financial support of the BAMP! project and this publication.

Graphic design, layout and typesetting Miriam Bussmann

Translation into English Usch Engelmann

Cover Paper Log House, Shigeru Ban, Sherman Contemporary Art Foundation, Sydney, 2017; photograph: Brett Boardman

Copy editing and project management Ria Stein

Production Anja Haering

Paper Magno Natural, 120g/m²

Printing Grafisches Centrum Cuno GmbH & Co. KG

Lithography Repromayer, Reutlingen

Library of Congress Control Number: 2022944000

Bibliographic information published by the German National Library
The German National Library lists this publication in the Deutsche Nationalbibliografie; detailed bibliographic data are available on the Internet at http://dnb.dnb.de.

ISBN 978-3-0356-2153-2
e-ISBN (PDF) 978-3-0356-2166-2

This book is also available in a German-language edition with the title *Bauen mit Papier*, print-ISBN 978-3-0356-2139-6.

P.O. Box 44, 4009 Basel, Switzerland
Part of Walter de Gruyter GmbH, Berlin/Boston

9 8 7 6 5 4 3 2 1

CONTENTS

Pavilions

Bridges

Interiors and Furniture

APPENDIX

1 PAPER IN ARCHITECTURE

Why build with paper? A material that usually fulfils completely different purposes, first and foremost the permanent documentation of knowledge in the form of a book, newspaper or written document, but also in hygiene and the packaging industry. This book will provide a detailed look at the reasons for building with paper.

Paper is made from wood. Humans have been using this renewable raw material for buildings since time immemorial. The material is currently experiencing a surge in development in many new applications and joining technologies that do justice to its anisotropic character. Because this is precisely where the material's potential lies: paper can be seen as an evolution of the basic material wood – a kind of Wood 2.0!

Wood is anisotropic, which means that it has different strengths in the x, y and z directions due to its natural growth. It is also characterised by imperfections in the areas of branching and by pre-existing damage due to the long growth process. Over the centuries, the structural handling of wood in terms of connection techniques and safety scenarios has developed in such a way that we have sufficient experience and knowledge to use the material safely and according to plan. At the same time, its uneven appearance shows it to be a natural, grown material. One example documenting the technological progress is laminated veneer lumber, developed to compensate for the irregularities of wood. Very thinly cut layers of wood (1 to 3mm) are glued on top of each other to form an engineered timber product in which the overlapping cancels out the imperfections of the wood. By rotating the layers in relation to each other, areas of different strengths can be balanced to achieve equal load-bearing capacity in at least two of the three directions. Another advantage is that the dimensions of the material, which are limited by natural growth, can be exceeded many times over: whereas boards cut directly from trees are limited in width and length, laminated veneer lumber boards are currently offered with maximum dimensions of 3.5 × 25m, a limitation that is due more to transport and machine design than to technical possibilities. The result is a homogeneous and largely natural material with dimensions that do not occur in nature and with very dimensionally accurate industrial reproducibility.

And now paper? Paper is made by separating wood fibres mechanically or chemically and then recombining these wood fibres in the form of thin flat layers. In addition, filler material or additives can be used to create other volumes and functionalities. In a way, paper production can thus be compared with the production process of laminated veneer lumber: disassembling and then reassembling creates a new homogeneous, industrially reproducible material that is available in varying dimensions. However, in the construction industry, paper has so far only been used in a few areas, for example, as separating or support layers and as experimental material. This is due to two main influencing factors: humidity and fire.

Wood and the derived timber products described above can absorb and release moisture without immediate damage to the material. With appropriate building-component dimensions, the risk of moisture damage can be further reduced. As far as fire protection is concerned, wood as a building and construction material can protect itself by charring the outer layers during the burning process and thus significantly delaying complete destruction by fire with loss of integrity.

This natural protective principle also applies to paper: by increasing the layer thickness, the burning process can be slowed down – however, at the cost of a heavy construction with more material. Moisture protection, on the other hand, is much more difficult to achieve. Recurring switches between humidification and dehumidification damage the paper structure in such a way that dimensional stability is no longer a given. Everyone knows the phenomenon: if a thin sheet of paper gets damp, it becomes wavy and remains permanently deformed. However, manufacturing processes have been developed that include the use of certain additives to the material to increase the wet strength of paper: a particular benefit in the fields of hygiene and packaging. This added property creates a potential that needs to be translated for applications in the building industry.

One unbeatable advantage of the material paper is the possibility of chemically dissolving the fibre structure of the base material wood and reassembling it. As mentioned above, this process offers reproducibility and material perfection, which, in turn, positively affects dimensioning and geometric diversity. Layering or shaping the material creates three-dimensional geometries such as corrugated board or honeycomb board. These structures enable lightweight yet structurally efficient sandwich building components that can themselves be layered and designed according to specific load requirements.

The development of building components can build on experience gained in the packaging and transport industry. The following three aspects, in particular, offer great potential for functional transferability to the building industry:

- techniques for joining paperboard or cardboard;
- layering of load-bearing and volume-generating components; as well as
- superimposition of the individual levels to avoid anisotropy.

In parallel, it is possible to build on the significantly improved paper manufacturing process itself, which has been consistently optimised over the past few decades. The production processes, which used to be very energy- and water-intensive, have been significantly improved through research and development. The same applies to circular economy: paper recycling stood at the beginning of all recycling efforts; it has long since been comprehensively established, and, technologically, it is mature.

Research project "BAMP! – Building with paper"

The fundamental research project "BAMP! – Building with paper", a programme funded by the State of Hesse, Germany, formed the starting point of this book. The interdisciplinary composition of the participating scientists made it possible to map and investigate the entire value chain of a building made of paper **»fig. 1**.

Building with paper is still in its infancy stage, but several experiments already promise design and functional opportunities. And there are ecological advantages: the recyclability of paper conserves resources, and the use of wood as a raw material reduces CO_2 emissions. All this could help to further develop the idea of ecological,

1 The research project "BAMP! – Building with paper" examines and analyses the entire value chain of a building made of paper.

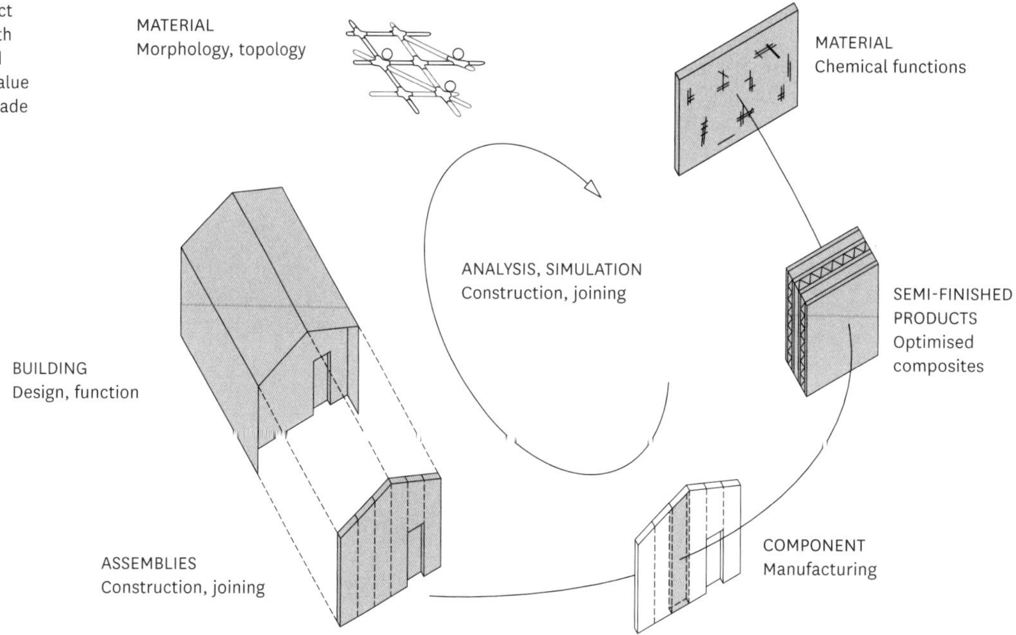

sustainable architecture – especially against the background of a material-related circular economy – which other building materials cannot offer in the same form.

Developing high-performing building systems made of wood has been a long journey. In order to establish building with paper as an independent construction method, it is necessary to examine to what extent the proven techniques and approaches of other construction methods can be transferred and where completely new paths need to be taken. Many questions are still open and, due to the complexity of the issue, they can only be solved jointly: thus, not only the industry but also designers, users, owners and operators of buildings are called upon to recognise the potential, develop it together and make it permanent.

To identify industrial solutions in the building industry, missing basics such as material parameters must be worked out and verified depending on the area of application. What requirements do buildings and components impose on the material and its manufacture? What standards result from the material and its use, and where can it be used for what purpose? In parallel, it will be necessary to investigate how existing materials from the paper industry can be used in architecture, how technologies from paper production can be transferred and to what extent they can be adapted for building objectives.

In a further step, these findings can then be translated into constructions. Building with paper can be classified as lightweight construction; it can be used for temporary structures and will develop its own design and functional patterns. It may be possible to use existing products and improve them (optimisation) intelligently or to develop other products with new methods (innovation) that are geared to specific applications in the construction industry. However, such pioneering work requires a high degree of interest, creativity and willingness on the part of all those involved – industry, designers and users » **fig. 2**. Only if the topic attracts sufficient attention can grants and funding be obtained for application-oriented basic research. And to gain acceptance in society, it must be clarified whether new framework conditions can be created

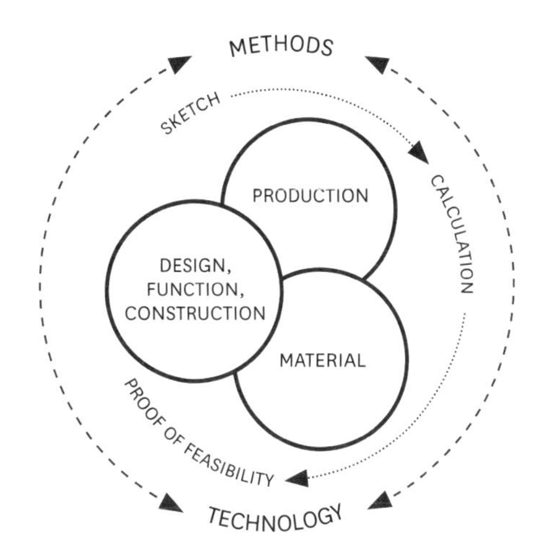

2 Development method of the research project "BAMP! – Building with paper".

for the material. Thus, more impetuses from designers and manufacturers lead to more projects and, in turn, to more points of contact. Standards and laws can prove critical for certain areas of application: requirements and regulations for temporary buildings for residential and recreational use need to be reconsidered and rethought to allow for alternative concepts with paper to be implemented in practice.

Research and development, also the subject of this book, help define the material's limits, including areas it is unsuitable for. Although many aspects of building with paper are still incalculable in terms of regulations, building standards and specific technologies, it is by no means utopian to conceive of building houses with or even entirely from paper in the next few years. The following chapters present aspects and impulses for further developing building with paper.

The history of paper in architecture

The origins of building with paper lie in ancient China and Japan. The invention of paper in 105 AD is attributed to the Chinese Minister of Agriculture Tsai-Lun. Mulberry, flax, silk or hemp fibres, but also old rags, were used to make paper; the mixture was pulped and mixed with mucilaginous substances in a water solution. The resulting pulp was poured onto screens, pressed and dried in the sun.[1]

The first paper applications in architecture were free-standing room dividers made of single or multiple panels. Such folding walls consisted of a paper-covered wooden frame; the paper was often painted. Later, the free-standing room dividers were further developed into sliding wall panels. Japanese architecture knows two variants: the translucent wall panels called Shoji consist of a wooden lattice frame covered with Washi paper. They allow daylight to pass through the wafer-thin paper and diffuse light illuminates the house's interior. The Fusuma panel also consists of a wooden lattice frame covered with paper, but the paper is opaque and often painted » **fig. 3**.

This traditional Japanese construction method with wood and paper creates hardly any sound insulation and therefore requires quiet and self-restrained behaviour from the residents.[2]

In Europe, the paper manufacturers of Córdoba, Seville and Valencia in Spain laid the foundations for paper production in the 13th century. Europeans used paper primarily as an information carrier and packaging material from the beginning. It was made

3 Japanese culture knows two panel variants: the semi-transparent Shoji (left) and the opaque, painted Fusuma (back). The paper partitions shown are in the Nazen-ji temple in Kyoto.

from rags and fabric scraps, and manual production proved difficult and expensive. The discovery of wood as a new raw material for paper production was preceded by the observation of wasp and hornet nests in the 19th century; such nests consist of chewed wood fibres. This discovery and the paper machine invented in 1799 by French inventor Louis-Nicolas Robert, which could produce paper in continuous strips, led to faster manufacturing processes and lower costs, paving the way for the spread of paper across the continent.

Other inventions – such as the first machine for the production of two-ply paperboard by John Dickson (1817), a folded paper by Edward Haley and Edward Allen (1856) and corrugated board with two cover layers by Olivier Long (1874) – opened up new possibilities in the application of paper and cardboard in the packaging industry. These developments were followed by new methods of producing hard-wearing papers (e.g. wrapping paper in 1879) and the development of new products such as cardboard tubes (beginning of the 20th century) or honeycomb board (around 1940). With advancing technology, material properties such as water resistance, fire protection and resistance to fungal infestation also improved.[3]

First houses made of paper

Inspired by the innovations in the paper and packaging industry, the first attempts to implement paper products in architecture occurred in Europe as early as the second half of the 19th century. French company ADT invented a prefabricated summer house, a hospital and a house for tropical climates and presented them at the World's Fair in Paris in 1889 »**fig. 4**. The prefabricated building components of these structures consisted of a double layer of 4mm thick cardboard on both sides of a 100mm thick cavity formed with U-shaped spacers. These 3m high and 600 to 800mm wide panels were easy to transport and assemble.[4]

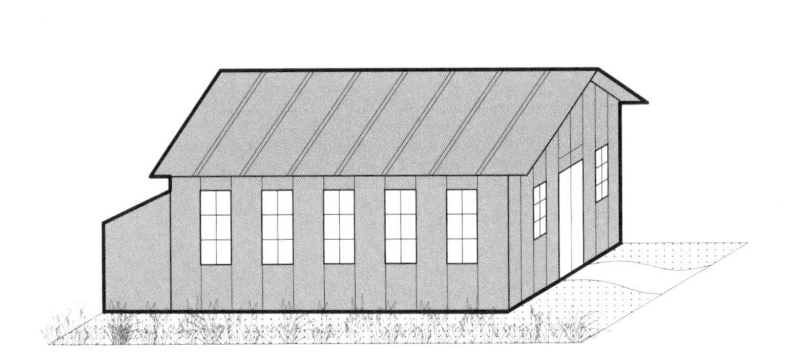

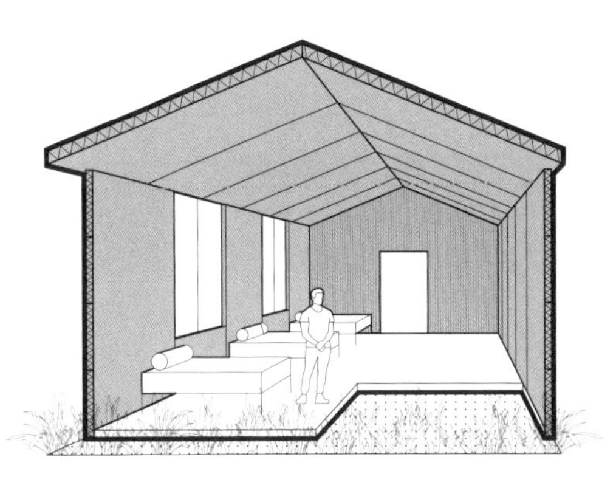

4 First documented cardboard house made of prefabricated elements for warm climates, Paris, 1889 (reconstruction).

The housing shortage after the Second World War was another reason for research into inexpensive and easy-to-assemble housing units. In 1944, the American Institute of Paper Chemistry developed an experimental design of small, transportable and expandable emergency shelters. The 2.4 × 4.8m units consisted of 25mm thick prefabricated corrugated cardboard panels made from waste paper. To impregnate the cardboard, the material was first soaked in sulphur and then covered with several layers of fireproof paint. Although these emergency shelters were designed for a one-year lifespan, individual specimens lasted 25 years.[5]

Due to their low weight, low production costs and structural efficiency, cardboard products were suitable materials for support structures and formwork for a wide variety of construction methods. The best-known examples are the corrugated cardboard panels with aluminium sheet covering the dome-shaped building by Richard Buckminster Fuller and students of McGill University in Canada (1957) and the honeycomb board panels with aluminium covering the Bear Zone Houses in New Mexico by Steve Bear (1971). Well-known examples of the combination of paper materials with plastics are the dome-shaped house by Container Corporation of America (1954) and the cardboard panels laminated with polyurethane foam by the Architectural Research Laboratory of the University of Michigan in Ann Arbor (1962–1964). Keith Critchlow and Michael Ben-Eli used corrugated cardboard panels as formwork for thin, sprayed-on concrete layers with chicken wire reinforcement (1967).

In the temporary utility unit Pappeder, the 30mm thick corrugated board was not only used as a substrate for the fibreglass coating but also provided structural stability and served as thermal insulation for the entire system. The prefabricated units had a square floor plan of about 11m^2. They could be combined into larger units to create different spatial constellations. Pappeder was designed by 3H Design (Hübner + Huster) in 1970 » **figs. 5, 6**. A total of 89 of these units were entirely prefabricated, transported to the site on low-loaders and erected with cranes on previously prepared foundations on the Olympic sites in Munich and Kiel in 1972. The units served as recreation and changing rooms, kitchenettes, first-aid rooms and toilets.[6]

Another interesting example of temporary constructions was the Plydome (the name is made up of the two words plywood and dome). With its folded, anti-prismatic plate structure, this accommodation was inspired by the Japanese origami art of folding. The construction is based on a three-hinged frame that could be folded flat for

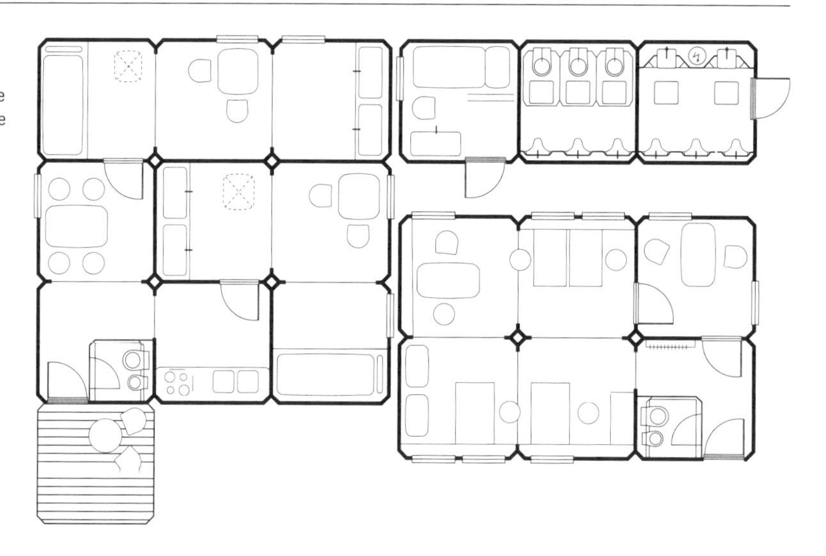

5 Floor plan Pappeder, 3H Design, 1970. The prefabricated units made of corrugated board were used, among other things, as changing rooms at the Munich Olympics.

6 Pappeder toilet unit, Munich, 1971.

transport. After delivery, two sections were joined together on site and extended to create a 5.8 x 5.2m structure with the highest point reaching 3m. The shell was anchored to the chipboard floor panel; then the side walls were attached. The sandwich panels with a polyurethane core and a 10mm thick solid cardboard layer on each side were coated with polyurethane to make them waterproof. Plydome, designed by Herbert Yates and developed with Sanford Hirshen and Sim van der Ryn in 1966, was a prefabricated housing kit that also included the main furniture for sleeping and storage **» figs. 7, 8**. Over a thousand Plydome units were made and used as housing for seasonal workers in California.[7]

In 1975, Dutch architect Paul Rohlfs designed an eco-house as part of his graduation project at Eindhoven University of Technology. The house was further developed from 1975 to 1980; several prototypes of the building envelope were created. The final prototype consisted of honeycomb sandwich panels with breathable foil applied on the outside and a vapour barrier on the inside. The corners of the construction were formed by special cutting and folding techniques that are used to fold sandwich panels. Two adjacent building elements were connected with folded flaps that were bolted together. The question was whether the construction would survive the winters in the province of Groningen. In fact, the unit was occupied for several years, which proved the good thermal quality of the building envelope and the resistance of the cardboard sandwich panels to adverse weather conditions.[8]

The influence of paper architect Shigeru Ban

The contemporary era of building with paper can be traced back to Japanese architect Shigeru Ban. Since previous research and projects were designed for short-term or temporary solutions, Ban, from the beginning, focused his interest on developing new solutions for implementation in experimental constructions. His enthusiasm for this construction method soon earned him a reputation as the "paper architect". Ban's fascination with paper, fed by the tradition of paper use in his homeland, first became apparent in the designs he developed for exhibitions on the two international architects Emilio Ambasz in 1985 and Alvar Aalto in 1986 at the Axis Gallery in Tokyo. In the

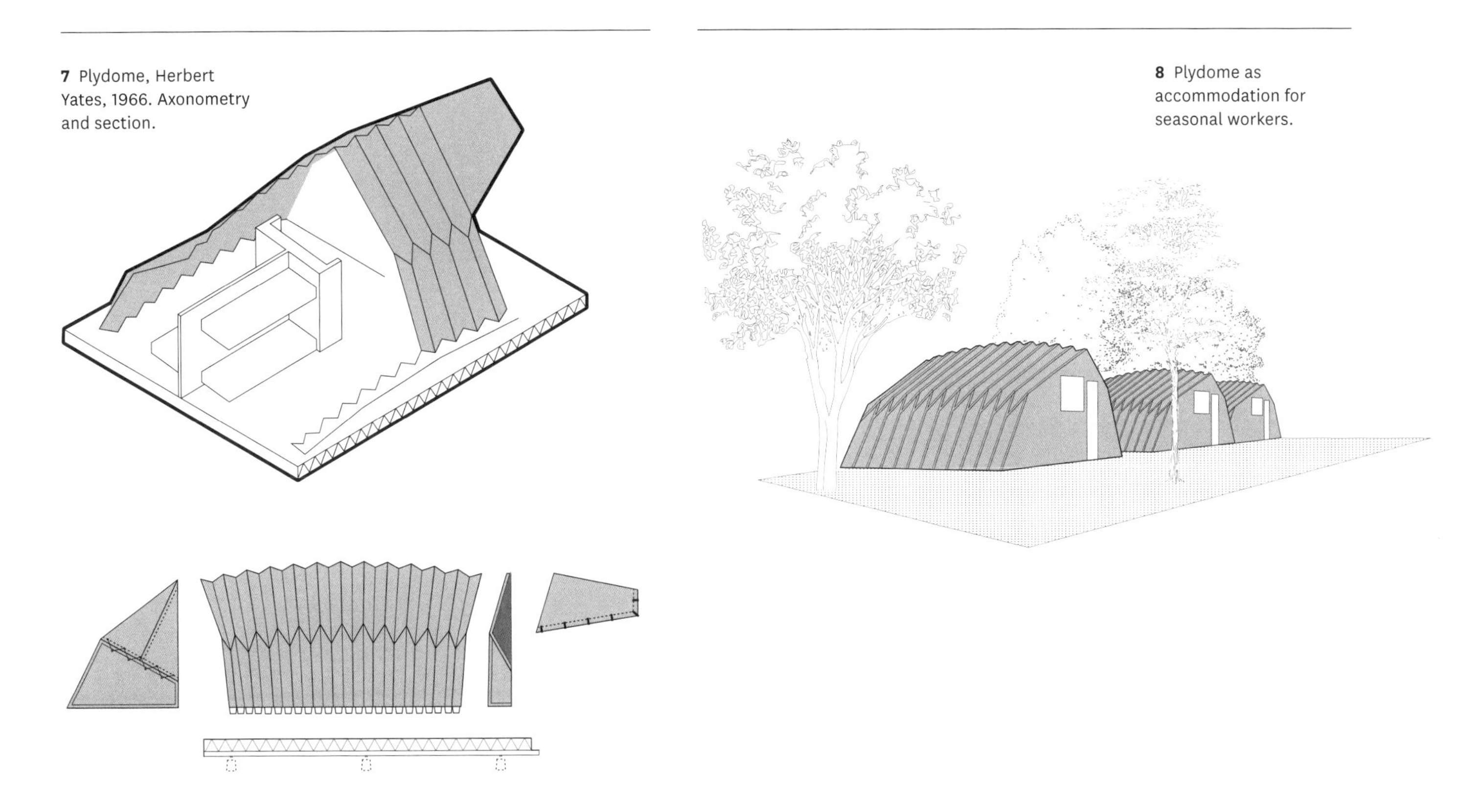

7 Plydome, Herbert Yates, 1966. Axonometry and section.

8 Plydome as accommodation for seasonal workers.

first project, Ban used square cardboard tubes and honeycomb boards to organise the space. For the Aalto exhibition, he actually wanted to use wood, one of the architect's favourite materials. Due to the limited budget, however, he opted for cardboard tubes as the main construction material for the entire exhibition **» fig. 9**.[9]

Ban once said that his work with paper was not initially based on its favourable ecological properties but that he generally disliked throwing things away. There were simply a lot of cardboard tubes left over in his office from using tracing paper. The interest in cardboard tubes as a building material arose when working on larger constructions. Ban began experimenting with the material, testing it for compressive and tensile forces and possible impregnation methods.

His first architectural construction to use cardboard tubes was Paper Arbour. This pavilion, designed for the World Design Expo in Nagoya in 1989, referred to the approach of Japanese gardens to create a contemplative space in quiet seclusion from the crowds of visitors and the noise of the fair. The pavilion consisted of 48 cardboard tubes arranged in a circle, each 4m high and stiffened with adhesive. The tubes' diameter was 330mm with a wall thickness of 15mm. They were made waterproof with paraffin impregnation, placed on a concrete foundation and connected at the top by a wooden tension ring. The textile roof that spans the construction was also attached to this tension ring. During the day, daylight could penetrate the inside of the pavilion through the gaps between the cardboard tubes; at night, they let the light inside the pavilion make it look like a glowing lantern. Next to the pavilion, an undulating partition wall with built-in seating was erected, also made of cardboard tubes.[10]

9 Alvar Aalto exhibition, Axis Gallery, Tokyo, 1986. The exhibition design by Shigeru Ban was based on cardboard tubes.

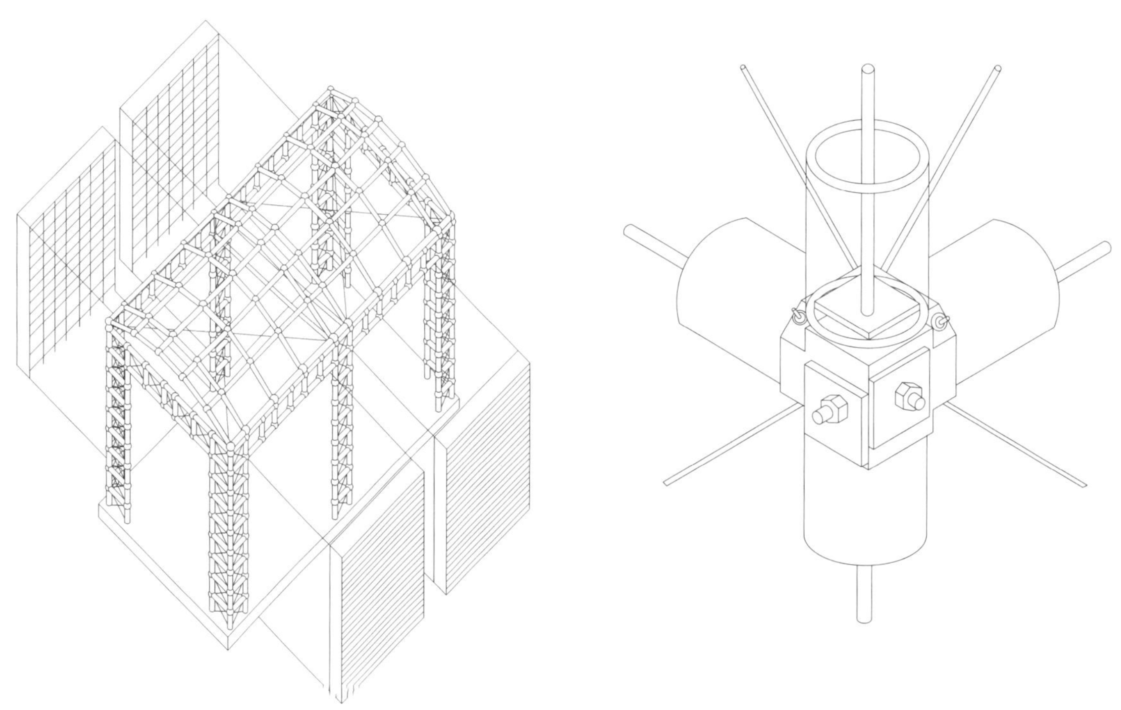

10, 11 Shigeru Ban, Library of a Poet, 1991. Axonometric and detail of the connection of the cardboard tubes.

12 Paper Log House, 2013: Demonstrator of the emergency shelters by Shigeru Ban.

Ban's breakthrough in paper architecture came in 1991 with the Library of a Poet – an annexe to the House for a Poet extension that enlarged and improved the existing building **»figs. 10, 11**. This library is the first permanent structure whose load-bearing elements consist mainly of paper. The vaulted roof rests on six support beams made of cardboard tubes. The walls are also made of cardboard tubes (ø 100mm, wall thickness 12.5mm) and are connected by wooden joints. Steel cables run through the interior to re-tension the construction, and diagonal braces are installed to stiffen it. Between the supports, four wooden bookshelves anchored to the floor increase rigidity against lateral wind forces. The roof construction consists of ten arches, also made of cardboard tubes connected with wooden joints, which are held down with the help of two cardboard trusses with internal, tensioned steel cables. The cardboard tubes are protected from the weather by the roof covering and glazed walls.[11]

Another important project created by Shigeru Ban was Paper House – a weekend house for the architect himself, who applied for a permit for it and obtained one in 1995 for a permanent structure based on cardboard tubes **»chapter 6, pp. 86–87**.

Also in 1995, the year of the Kobe earthquake, Ban and volunteers built low-cost log cabins out of cardboard. The 16m^2 Paper Log Houses **»fig. 12** were used as emergency shelters for Vietnamese refugees severely affected by the earthquake **»chapter 6, pp. 88–89**. The Takatori Paper Church was built in addition to the houses. This sacral building also consisted of a cardboard tube construction whose oval ground plan referred to the churches of the Baroque period and thus offered people a new religious home. In 1995, Ban founded the non-governmental organisation Voluntary Architects Network, which provides emergency shelters and other structures for victims of natural disasters and other calamities worldwide. The design of emergency shelters has

13 Shigeru Ban, Hualin Primary School, Sichuan, 2008, 1:1 model, 2013.

become a trademark of the Japanese architect, who was awarded the Pritzker Prize for it. Variants of the Paper Log Houses can be found in Turkey (2000), India (2001) and the Philippines (2014). In China, Ban worked with VAN to build the Hualin Primary School »**fig. 13** in the eastern part of the city of Chengdu in Sichuan Province in 2008: three buildings with a frame structure of cardboard tubes connected with wooden joints. His works also include the Miao Miao preschool in Taiping town, built in 2013, a post-and-beam construction based on cardboard. The Cardboard Cathedral, a replacement structure after the earthquake in Christchurch, New Zealand, consisted of 16m long cardboard tubes (ø 600mm) assembled to mimic the shape of a triangular nave. To accommodate the large span of the structure and to prevent bending, wooden logs were inserted into the cardboard tubes that formed the main supporting elements.

To date, Shigeru Ban has designed over 60 projects in which cardboard tubes serve as the main construction material. The largest cardboard construction ever built was his Japanese Pavilion for Expo 2000 in Hanover. Ban designed this 73.5m long, 25m wide and 15.9m high shell construction together with Frei Otto »**chapter 6, pp. 132–133**. Cardboard tube arch constructions are a concept that Ban implemented at various scales in several of his projects, for example, in the Paper Dome (1998), the Paper Studio at Keio University (2003), the Paper Temporary Studio on the roof of the Centre Georges Pompidou in Paris (2004) and the Shigeru Ban Studio at Kyoto University of Art and Design (2013) »**chapter 6, pp. 112–113**.

Further developments in paper architecture

In the 1990s, increasing attention was paid to the environmental risks posed by building products. More environmentally friendly products and materials were developed. One vital impulse for this trend was the Earth Summit, the United Nations Conference on

Environment and Development in Rio de Janeiro, in 1992. In the same year, Dutch architecture professor Hans Ruijssenaars designed the Apeldoorn Cardboard Theatre, which was almost completely recycled after six weeks of operation »**chapter 6, pp. 130–131**.

Since 2000, interest in paper and cardboard as renewable materials has increased further and research has begun to intensify. Westborough Primary School was designed by Cottrell & Vermeulen Architecture in collaboration with Buro Happold and built in Westcliff-on-Sea, UK, in 2001 »**chapter 6, pp. 96–99**. It is the first European example of a durable cardboard building designed for a 20-year lifespan. The aim was to construct a building whose main structure was made of cardboard tubes and honeycomb composite panels and which could be 90% recycled at the end of its lifespan. Another building in Europe with a similarly experimental approach is the Ring Pass Hockey and Tennis Club extension, designed and built by the Dutch engineering company Octatube in 2010 »**chapter 6, pp. 102–103**. This project's primary focus was on the spatial roof structure made of cardboard tubes and its associated impregnation methods. Currently, Fiction Factory, an Amsterdam-based company, produces prefabricated dwellings called Wikkelhouses. The modules of these "wrapped" houses are made of corrugated board that is wrapped around a mould, then stiffened with eco-friendly glue, and finally covered with wood on the inside and outside »**chapter 6, pp. 90–95**. The company guarantees that the houses will last at least 15 years but expects a lifespan of up to 50 years.

Apart from a number of successful permanent buildings, temporary structures are an obvious application for easily recyclable materials such as paper and cardboard. Depending on the intended function and service life, other materials such as aluminium, polyurethane coatings or polyurethane foam often complement the load-bearing structure in paper or cardboard architecture. Most of Shigeru Ban's projects were designed for temporary use of up to five years. However, some lasted longer or were transferred to permanent use. The IJburg Paper Theatre, designed by Ban in collaboration with Octatube, was set up twice: in 2003 in Amsterdam and in 2004 in Utrecht »**chapter 6, pp. 134–135**. It is currently in storage and will be rebuilt in 2023 in Bijlmermeer, Amsterdam Zuidoost, to serve as a youth centre.

Public Farm 1 by WORKac in New York was an architectural and urban planning manifesto from 2008 that consisted of huge cardboard tubes functioning as large temporary flower and herb containers arranged into mounds »**figs. 14, 15**. In another project, students at ETH Zurich, under the direction of Tom Pawlofsky, designed and built a circular pavilion composed of 409 cone sections made of 28-layer corrugated board. The cones were prefabricated and sent to a temporary exhibition in Shanghai in 2010. Students from Wrocław University of Science and Technology designed and built the Poet Pavilion in 2018. It was composed of a wooden foundation, cardboard tubes and a roof construction made of uncoated corrugated board, covered with polycarbonate and plywood panels. After nine months, the pavilion was dismantled, and the entire roof structure could be recycled.

The history of paper architecture shows that development runs in two parallel strands. In addition to permanent buildings, there are many experimental and temporary projects. These are often supported and accompanied by scientific research projects. Examples can be found at McGill University in Montreal in the 1950s; the Polytechnic of Central London and California Polytechnic State University in the 1970s; Waseda University, Chiba Polytechnic College and the Technical University of Dortmund in the 1990s, TU Delft and ETH Zurich in the first decade of the 21st century, and Wrocław University of Science and Technology (WUST) and TU Darmstadt in recent years.[19]

14, 15 Public Farm 1 by WORKac, New York, 2008. Paper tubes were used to build temporary large-scale flower and herb containers.

The long tradition of paper in architecture as well as the large number of research projects show a constant interest in further developing this construction method. Industries such as packaging, automotive or aerospace provide an impetus for improving paper products. Its basic material, cellulose, is renewable – and thus, paper, with its historical roots, has the potential to become an important future building material.

REFERENCES

1 Kiyofusa Narita [1954], *A life of Ts'ai Lung and Japanese paper-making*, Tokyo: The Paper Museum Tokyo, 1980; Józef Dąbrowski, Jadwiga Siniarska-Czaplicka, *Rękodzieło papiernicze*, Wyd. nakł. Wydawnictwa Czasopism i Książek Technicznych "SIGMA" NOT, Spółka z o.o, 1991.
2 Dianne van der Reyden, "Technology and Treatment of a Folding Screen: Comparison of Oriental and Western Techniques", in: *Studies in Conservation*, 1988, 33(1), pp. 64–68; Yuji and Chosuke Taki Kishikawa, All Japan Handmade Washi Association (eds.), *Handbook on the Art of Washi*, Tokyo: [illegible]
3 Christopher J. Biermann, *Handbook of Pulping and Papermaking*, San Diego: Academic Press, 1996; Stefan Jakucewicz, *Wstęp do papiernictwa*, Warsaw: Oficyna Wydawnicza Politechniki Warszawskiej, 2014; William E. Scott, James C. Abbott, Stanley Trosset, *Properties of Paper: An Introduction*, 2nd, revised edition, Atlanta: Tappi Press, 1995.
4 Gernot Minke, *Alternatives Bauen: Untersuchungen und Erfahrungen mit alternativen Baustoffen und Selbstbauweisen*, Kassel: Forschungslabor für Experimentelles Bauen, Gesamthochschule Kassel, 1980.
5 Roger Sheppard, Richard Threadgill, [illegible] *Survival Scrapbook 4*, Carmarthen: Unicorn, 1974.
6 Pamela Voigt, *Die Pionierphase des Bauens mit glasfaserverstärkten Kunststoffen (GFK) 1942 bis 1980*, Dissertation, Bauhaus-Universität Weimar, 2007.
7 Vinzenz Sedlak, "Paper Structures", in: 2nd International Conference on Space Structures, University of Surrey, 1975, pp. 780–793.
8 Mick Eekhout, Fons Verheijen, Ronald Visser, *Cardboard in Architecture*, Amsterdam: IOS Press, 2008.
9 Philip Jodidio, *Shigeru Ban: Complete Works 1985–2010*, Cologne: Taschen, 2010.
10 Riichi Miyake, Ian Luna, Lauren A. Gould, *Shigeru Ban: Paper in Architecture*, New York: Rizzoli International Publications, 2009.
11 Matilda McQuaid, *Shigeru Ban*, London: Phaidon, 2003.
12 Jerzy F. Latka, "Paper in Architecture: Research by Design, Engineering and Prototyping", in: *A+BE – Architecture and the Built Environment* (19) 2017, https://journals.open.tudelft.nl/abe/issue/view/547, accessed 19 March 2021.

2 MATERIAL

For a better understanding of the properties of paper, the following chapter provides deeper insights into the chemistry and structure of cellulose and the extraction of the pulp.

Paper is a planar material, mostly made of plant-based fibres, formed by removing the water from previously soaked fibres on a screen. According to DIN 6730 (Paper, Board and Pulps – Terms)[1] and DIN 6735,[2] paper is also defined by the compression and drying of the resulting fibre felt. The manufacturing process of paper thus differs significantly from that of woven textiles, dry-laid non-wovens, and similar aerodynamically or mechanically manufactured fibre roving.

As described in chapter 1, papermaking originated in China, and the material spread to Europe via the Mediterranean region. In this early phase, the core task of the paper mills spreading across the entire continent was preparing the raw paper material. Until the 19th century, this consisted exclusively of old textiles made of hemp, linen and cotton (rags) as well as rope-making and spinning waste. These materials were gathered by rag pickers and pounded into fibre pulp in the paper mills. As the increase in printed products drove the demand for paper considerably from the 18th century onwards, bottlenecks soon occurred in raw material supply.

In 1843, driven by the search for new sources of fibre for paper production, it was finally possible to convert the raw material wood into usable fibres by comminuting it mechanically. In this energy-intensive process known as wood grinding, tree trunks were frayed by pressing them against a wet grindstone. Although this technique produced a pulp (wood pulp), it did not meet all the requirements of papermakers and users. As a result, other ideas for pulping wood developed in the following decades. Particularly noteworthy in this context are the milestones of chemical wood pulping, such as the "sulphite process" developed in 1867 or the "sulphate process" first described in 1870. In these approaches, wood was not left in its original composition and simply comminuted but was also treated with suitable chemicals to dissolve the lignin out of the (natural fibre composite) wood under pressure and temperature. The pulp obtained in this way, consisting mainly of cellulose, is still the most important raw fibre material in the paper industry today.

Wood composition and pulping processes

Paper materials, therefore, consist mainly of fibrous materials obtained from the highly complex, biogenic fibre composite wood. Wood fibres have a multi-layered cell wall structure. As »**fig. 1** shows, the walls of wood cells are made up of several individual layers of varying thickness, interspersed with so-called fibrils. Put simply, wood consists of tensile cellulose fibres, a surrounding lignin matrix that provides hydrophobicity and compressive strength to the overall composite, and a class (i.e. a group of elements) of variably structured polysaccharides (polyose).

Mechanical and chemical wood pulping processes are used to overcome the adhesive forces of the lignin material to break down the wood into usable fibres.

Mechanical processes are used to obtain groundwood pulp or wood pulp. In terms of process technology, a distinction is made between stone grinding and refining. In both processes, the fibres are frayed by mechanical energy with the help of moisture and temperature. The yield of this process is very high at around 96% of the starting material, as no substances of content of the wood are removed apart from a small proportion of water-soluble substances. In contrast to chemically obtained fibrous materials, mechanical wood pulp, therefore, still contains lignin. This can be recognised by the darker colour and lower flexibility, which results in a coarser structure overall. The lignin makes it more difficult for the fibres to bond in the paper » **following section "Cellulose and structure"**. Mechanical wood pulp is, therefore, not suitable for all applications.

In contrast, chemical pulping processes are used to dissolve the lignin out of the wood structure, which leads to significantly purer pulps. For this purpose, wood (mainly in the form of wood chips) is cooked under pressure and temperature by adding suitable chemicals. A distinction is made between sulphite and sulphate or kraft processes. Chemical digestion does not damage and fragment the fibres as much as mechanical processes. Thus, the fibres themselves are more stable and retain higher strengths. Since the lignin is almost completely removed, the resulting stronger fibre bonds yield strong paper material. The strength of paper depends essentially on the strength of the cellulose fibres themselves and on the bonds between the fibres.

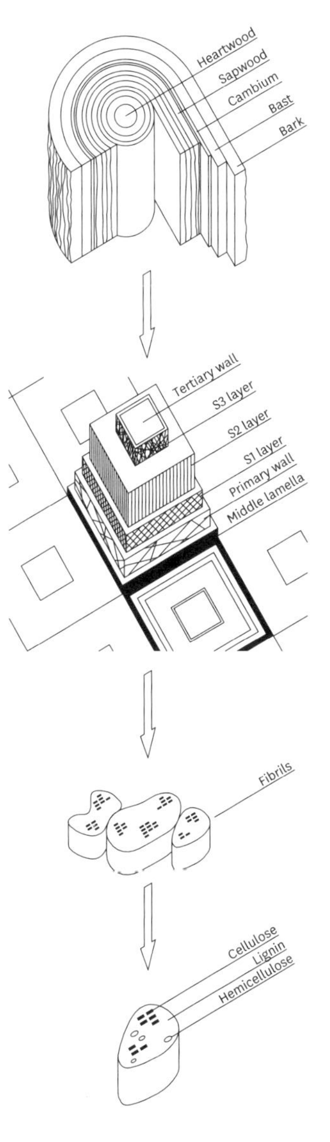

1 Hierarchical cell wall structure of a wood cell and its integration into the structure.

Cellulose and structure

Quantitatively, cellulose is considered the commonest biopolymer in the world. In nature, plants of all genera form an estimated $1{,}5 \times 10^{12}$ tonnes of cellulose annually.[3] In wood, the cellulose content reaches 40 to 47%, in the bast fibres of hemp, flax and jute around 70%, and in the seed fibres of cotton it even reaches more than 94%.[4]

Chemically, cellulose is a polymer: a long chain of linked glucose molecules (D-glucose). The molecular formula of the cellulose polymer is $C_{6n}H_{10n+2}O_{5n+1}$, where "n" stands for the degree of polymerisation (DP) of the cellulose » **fig. 2**. This degree of polymerisation, i.e. the length and structure of the cellulose molecule, is decisive for the strength of the fibre itself. In addition, there are hydroxyl groups (-OH) » **fig. 3**. On the one hand, these lead to intramolecular hydrogen bonds (red), which increases the stability of the cellulose molecule itself. On the other hand, the hydroxyl groups are decisive for the bonds between the cellulose fibres, i.e. for the formation of intermolecular hydrogen bonds (blue). The number of these hydrogen bonds between the individual molecules and the totality of the hydrogen bonds determines the strength of paper.[5]

The presence of lignin makes the formation of intermolecular hydrogen bonds more difficult. Therefore, pulps with the lowest possible lignin content are preferred for paper production.

Paper production

In Germany, 23,123 metric kilotons (kt) of paper were produced in 2021.[6] The quantity is distributed over four uses

2 Schematic of the molecular structure of a cellulose chain.

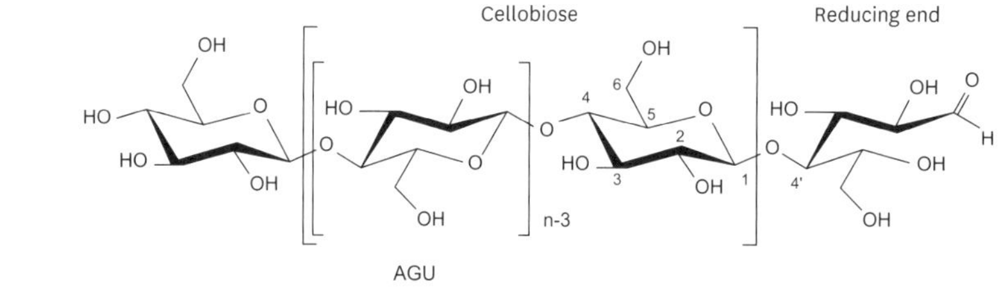

3 Hydrogen bonds within a cellulose chain (red) and between different chains (blue), which are dissolved when water penetrates, resulting in a loss of mechanical properties.

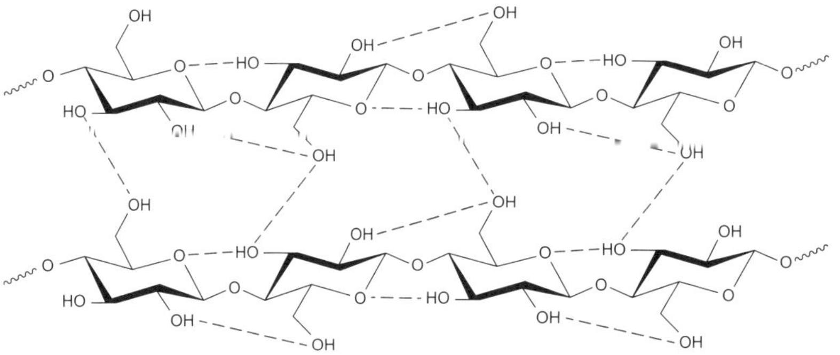

- 59% paper for the packaging industry,
- 28% repro paper,
- 6% hygiene paper,
- 6% papers for technical purposes and special applications.

In 2020, Germany was the largest producer of paper, paperboard and cardboard in Europe with 21,348 metric kilotons (kt). In a global comparison, Germany lay behind China (103,994kt), the USA (67,959kt) and Japan (22,887kt).[7]

As already mentioned, paper is mainly composed of fibres. These are present either as primary fibres, i.e. pulp or mechanical pulp obtained from fresh wood » **section "Wood composition and pulping processes", pp. 22–23** or as secondary fibres obtained from waste paper. The share of recovered paper fibres in the total mix was 79% in all of Germany in 2021. The selection of raw materials and the design of the papermaking machines were adapted to the products. The highest use rates of recovered paper are for corrugated paper (106%), paper for packaging (100%) and newsprint (113%). Recovered paper utilisation rates above 100% are necessary to compensate for losses caused, for example, by impurities that are removed in the preparation process » **section "Stock preparation", p. 26**. In contrast, the recovered paper utilisation rate for technical papers and special uses, including coreboard, was only 52%.[8]

Other raw materials include mineral filler materials such as calcium carbonate, kaolin, talc or titanium dioxide and chemical additives such as binders, sizing agents, fire retardants, and wet-strength and dry-strength agents. While fillers are always added to the fibre suspension, additives can be either added to the suspension or they can be applied as a coating to the surface of the finished paper, depending on their function. The latter process can be integrated into the paper production (in-line) or carried out separately in a downstream process (off-line). In the case of chemical auxiliaries, a

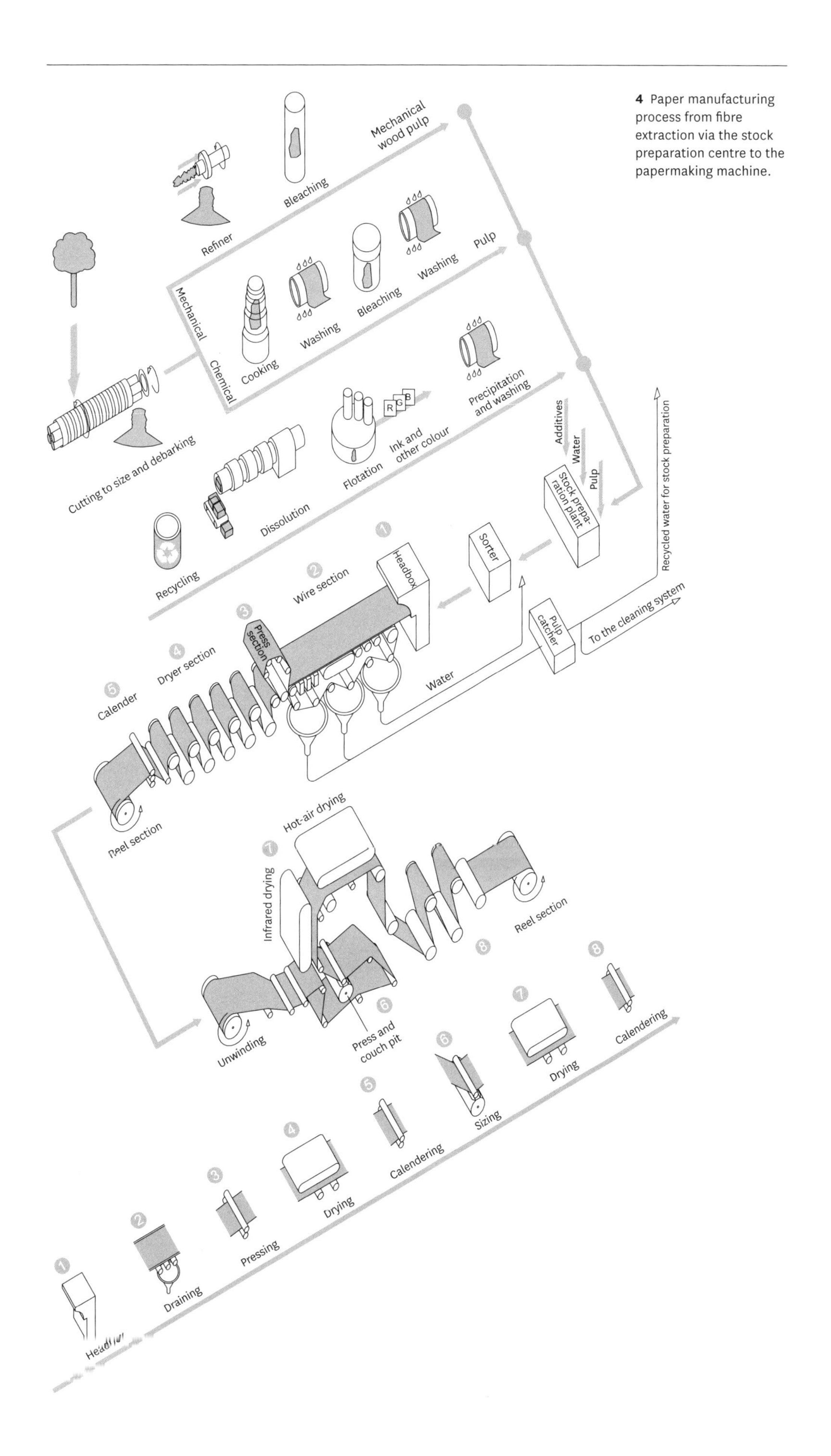

4 Paper manufacturing process from fibre extraction via the stock preparation centre to the papermaking machine.

fundamental distinction must be made between process chemicals and additives or functional chemicals. Process chemicals support the production process but have no direct influence on the paper properties and remain in the paper only to a small extent. Additives or functional chemicals, on the other hand, remain in the paper. After all, they are supposed to improve its properties. In the case of copying paper, the typical amount of filler material used is around 20%, based on the oven-dry weight of the paper. In contrast, additives are marginal, and often fewer than 1% remain in the paper.

Papermaking requires a lot of water **»fig. 4**. There are various water circuits within a paper mill so that the clean and the dirty water can be used economically, and the overall consumption can be kept as low as possible. In modern mills, the production of one kg of paper requires an average of 2–5l of freshwater (compared with 10l in older plants). Such water pollution is mainly due to the use of waste paper: during dewatering, fibre residues, filler materials and additives are flushed out. Factory-based treatment plants recycle the polluted water before it is discharged into canals, streams or rivers.

Stock preparation

The manufacturing process starts with the preparation of the fibres. These are often delivered as pressed bales because the pulping processes, described in **»section "Wood composition and pulping processes", p. 22**, are usually carried out at other sites. The bales are dissolved in the pulper in the stock preparation plant by being stirred in water until a suspension of water and fibres is obtained.

The individual work steps carried out in a stock preparation are oriented towards the end product. With regard to strength properties and paper construction, one crucial factor during stock preparation is grinding the fibres in the refiner. In the refiner, the fibres are mechanically treated so that they "fibrillate", i.e. the fibrils are detached from the fibre wall **»fig. 1**. This process increases the bonding surface between the fibres and thus boosts the strength of the finished paper. Treating the fibres in this way also facilitates the bonding of chemical additives.

Shortly before the prepared stock is fed to the papermaking machine, water is added to dilute it to about 1% stock consistency (ratio of solids to water) in the stock preparation plant. This production area is also called the approach flow section because the papermaking machine, unlike the pulper, for example, is operated continuously.

Papermaking machine

Depending on the type, papermaking machines can be over 10m wide and 120m long.[9] They are divided into three sections: the wet end or forming section, consisting of headbox, wire and press section; the dryer section; and the reel section. The machines can achieve production speeds of up to 2000m/min. The papermaking machine distributes the fibres evenly, produces a continuous paper web, dewaters it, dries it and finally winds it up on a roll.

Wet end

The wet end or forming section is the first station of the papermaking machine. At the headbox, the fibre slurry is evenly distributed over the full machine width and brought to the same speed as the papermaking machine (machine speed). It then passes onto the wire, where it is dewatered using gravity, foils (static elements under the moving wire creating a negative pressure) and vacuum. The paper should now present a good formation, i.e. be as homogeneous and flocculation-free as possible. Different machines are adapted to parameters such as production speed or grammage of the end product. A classic former is the Fourdrinier wire **»fig. 4**.

For multi-ply paperboard or cardboard, the machine may be equipped with a headbox and a wire for each ply. The still wet paper webs are pressed together ("couching"). At the end of the wire section, the paper web has a dry content of about 20%.

Press rolls then carry out further mechanical dewatering in the press section. The paper web is fed through a nip between two rollers. Two press felts, running above and below the paper web, absorb the pressed-out water. The decisive factors are surface area in the nip or dwell time and pressure. The press has a significant influence on the density as well as the strength and surface properties of the paper. At the end of the press section, the dry content has increased to around 50%; further drying takes place thermally in the dryer section.

Dryer section

In the dryer section, the paper sheet runs around internally steam-heated cylinders, pressed on by dryer screens. Thermal energy is introduced into the paper and the evaporating water is removed. The dryer section is the longest part of a papermaking machine. It is important to use the least amount of energy required during these processes and keep the process chambers as small as possible. The entire dryer section is enclosed to save energy, and any residual heat is recovered via heat exchangers.

Paper shrinks when drying, which is undesirable. This behaviour is counteracted by the tensions generated in the paper sheet and the pressing against the cylinders. This process, in turn, influences the mechanical properties of the paper: shrinkage-restrained dried paper is stronger than freely dried paper.

After the dryer section, only a little residual moisture remains in the paper. Paper is hygroscopic, which is why an equilibrium is established with the ambient air humidity. Accordingly, dry contents of about 93% are targeted.

Other in-line aggregates

Additional in-line units can be installed between the dryer and reel sections. For example, a sizing section enables the application of coatings to improve the surface properties visually or functionally. After the sizing is applied, the sheet must be dried without contact. This task is performed by infrared dryers, which guide the sheet through a stream of air.

A so-called calender ensures the paper's density, gloss and smoothness. The principle of operation is similar to that of ironing: the paper is pressed between rollers, with different combinations of rollers bringing about the desired properties.

Reel section

The last station is the reel section. The paper is rolled onto spools and then transported for further processing. This process is no easy matter: the spools can reach a total mass of up to [illegible] tonnes and have to be replaced on the fly as production continues.

Further processing

The previously described processes of sizing and calendering can also be installed as downstream processing steps. In this case, the paper is unrolled from the reel spool, then sized or calendered and finally rolled back onto a reel spool.

From the reel spool, the paper finally reaches the winder section. Since the edge area cannot be used, the edges of the sheets are trimmed. This waste is fed back into the cycle in the stock preparation system. In addition to edge trimming, the paper is cut to the width required by the customer and finally rolled onto paper tubes. If the customer prefers cut-to-size goods instead, sheets are cut in the sheeter and then palletised.

The last station is the shipping department, where the rolls and pallets are packed ready for shipping and wrapped in film, for example. Modern warehousing systems and transport techniques support further logistics.

Material properties

When building with paper, the focus is on its mechanical properties, behaviour towards moisture or water, gas permeability, flammability and resistance to harmful biological factors. Properties such as gloss, smoothness, roughness, brightness, printability and softness or haptics play a subordinate role and are not considered further here.

The strength of the material is the essential factor when considering the structural integrity of a paper building. Starting from a fibre distribution in the paper that is as even as possible, the fibre orientation has a decisive influence on the strength of the final product. In industrially produced paper, the fibres are aligned in the running direction of the machine. This fibre orientation depends mainly on the velocity difference between the white water stream (jet) and the screen. Thus, when describing the strength properties, a distinction is usually made between machine direction (MD, x-direction) and cross direction (CD, y-direction) » **fig. 5**. The fibres lie mainly in the x-y plane and form a three-dimensional network. The material behaviour in the thickness direction (z-direction) differs significantly, since with thin papers only a few fibre cross-sections lie on top of each other. Standard printer paper with a grammage of 80g/m² has a "thickness" of about 100µm (micrometres). However, the typical characteristic of a paper is not its thickness but its grammage. Paper shows auxetic material behaviour, i.e. it becomes thicker under tensile load.

Paper properties change with the web width (CD), as the edge area of the paper web can dry faster and is less shrinkage-restrained. Thus, sheets from the same machine taken at different positions can have different property profiles. The type of fibre or its morphological aspects, the degree of fibre grinding and the presence of fillers and additives have a decisive influence on the mechanical properties of the resulting paper. Long-fibre (kraft) pulps with an average fibre length of approx. 3mm from coniferous woods produce particularly strong paper.

The most extensively discussed factor in the literature is the relevance of hydrogen bonds to paper strength. If unmodified paper becomes moist, its hydrogen bonds separate and the individual fibres detach from each other – the structure disintegrates. Unmodified paper is hydrophilic and always absorbs a certain amount of moisture from the environment. This process is accompanied by a slight, directional increase in volume. The individual fibres swell radially by 10–40% when moistened but change only minimally in the longitudinal direction. This anisotropic behaviour of the fibre is transferred to the entire fibre web. Consequently, the moisture expansion in CD will always

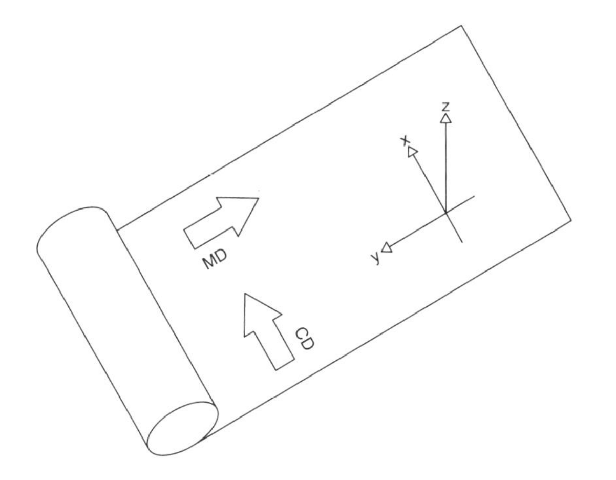

5 The axial directions in paper: MD = machine direction, CD = cross direction. The arrangement of the paper fibres is largely in MD due to the running direction of the papermaking machine, which is why the tensile strengths are usually greater in this direction.

greater than that for MD, a fact that must be taken into account regarding dimensional stability in structural applications and also in bonding with other materials. These processes are reversible through drying or at reduced humidity.

In a humid climate not only do the dimensions and strength change but other adverse effects can also occur. Suppose paper is combined with foreign materials such as cementitious materials, water glass (sodium silicate) or even base metals (which can lead to an acidic pH value). In that case, the material can become brittle when moistened.

It is much more critical if microorganisms colonise moist, unprotected paper, a process that is almost always accompanied by material damage. Since paper consists mainly of cellulose, previously isolated from plant material, it is not surprising that soil microorganisms, in particular, have highly specialised mechanisms to metabolise paper. This fact represents a serious advantage over other building materials regarding biocompatibility. But it creates a fundamental challenge when paper is assigned a structural function. Ground contact and direct weathering, in particular, can significantly reduce durability.

The flammability of paper also reduces its strength. At temperatures above 180°C, it begins to decompose without visibly melting. If there is enough oxygen present, a complete conversion to water and carbon dioxide occurs, leaving only a very small amount of mineral ash. In the case of papers containing fillers and additives, the combustion products that ensue may be chemically different. The fact that paper ignites easily is partly due to the high surface reactivity and the porosity of the fibre web. For structural applications, this means that fire protection precautions are mandatory.

Paper's chemical and biological properties depend on its respective ingredients (type of fibre, fillers, additives, contaminants) and cannot be described universally. In principle, paper is considered to be extremely robust, although pure cellulose papers have the limitations listed above (susceptibility to moisture and microorganisms, flammability). As the following explanations show, such influences can be countered with additives or appropriate downstream process steps that offer functional optimisations such as sizing.

Contact with water

The interaction between paper and water is mainly influenced by two factors: the fibre surface's chemical nature and the paper's microstructure with its pore volume and capillaries. If paper comes into contact with water, the surface is wetted first before the water penetrates the paper surface and spreads evenly over it. From the surface, water can penetrate the paper not only via the capillaries – the spaces between the fibres – but also through the open structure of the fibres themselves. The water molecules penetrating the paper reduce the strength of the hydrogen bonds, and thus the paper fibres themselves until the paper eventually disintegrates completely.

Wet strength

Once the hydrogen bonds have dissolved, the wet strength of the paper is only 3 to 10% of its dry strength. So-called wet-strength agents increase strength by cross linking the paper fibres with water-resistant bonds. They form stable covalent bonds at the fibre-fibre intersections, which cross-link the fibres permanently and thus also in the wet state »**fig. 6**. Wet-strength agents are used for papers that must remain stable when in contact with water, such as kitchen rolls, tea bag papers or filter papers. These properties are of great relevance for structural applications, as well. By definition, wet-strength papers must have a relative wet strength of 15% of the dry strength, which can be determined using tensile tests according to DIN 54514.[12]

Classic wet strength agents are resins based on polyamidoamine epichlorohydrin (PAAE), melamine-formaldehyde or urea-formaldehyde. There are also temporary wet-strength agents, for example, for toilet paper production. After use, the paper should lose its wet strength so that it does not clog the sewage system. The substances used to achieve this behaviour, primarily multi-functional aldehydes (for example, glyoxal), can be reversibly cleaved.[13]

Therefore, wet-strength agents are added to many papers to provide residual stability when wet. However, they do not prevent water from penetrating the paper structure.

Sizing of paper

If the penetration of a liquid is to be temporarily prevented or slowed down, sizing must impart water-repellent (hydrophobic) properties to the hydrophilic paper or cellulose fibres. This reduces the wetting of the paper surface and inhibits the penetration of water into the paper. Basically, sizing can be done in two ways:

- In internal sizing, the sizing agent is added to the pulp before sheet formation.
- In surface sizing, the sizing agent is applied to one or both sides of the dry paper sheet.

Suitable sizing agents are resin glues, neutral sizing agents such as alkyl ketene dimers (AKD) and alkyl succinic anhydrides (ASA) or polymeric sizing agent systems (PLM). The effect of the sizing agent is determined by identifying the water absorption capacity according to the Cobb method (DIN EN ISO 535:2014-06)[14] or by determining the wettability via contact angle measurements.

Although sizing makes the paper surface hydrophobic, it does not remain entirely and permanently water repellent. However, water repellence can be achieved with

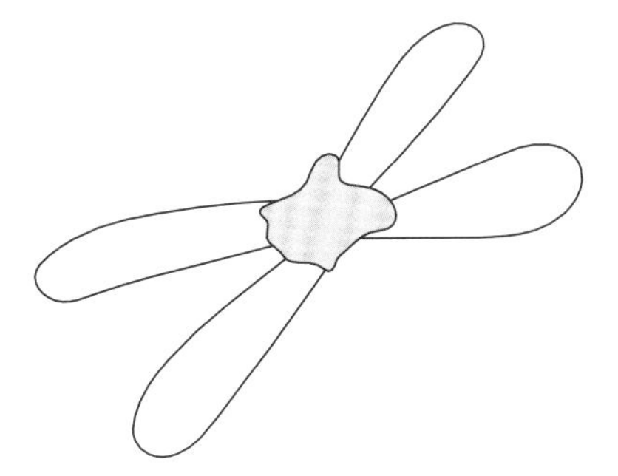

6 Cross-linking of two paper fibre intersections (white) through firm bonding of a wet-strength agent (grey).

other techniques: one common method is surface lamination with plastics such as PE, PET or PLA. However, for building with paper, the ecological framework conditions (material composition, recyclability, degradability) must also be considered. Coating materials are required that are effective even in small quantities and that biodegrade in a similar manner to the substrate paper itself. They must, however, have no negative consequences on the recyclability of the paper. One promising approach is superhydrophobic papers with the so-called lotus effect, named after the leaf of the lotus flower with its extremely water-repellent properties.[15] This effect is due to the micro- and nanoscopic architecture on the leaf's surface. To imitate it, many research approaches pursue coatings with hydrophobic nanoparticles such as functionalised silica[16] or cellulose nanoparticles.[17] To date, however, none of these superhydrophobisation strategies have been established in the papermaking process, as they do not meet all the requirements of paper producers and users. Either there is a lack of material safety, the application technologies are too complicated or the required recyclability is not met. There are various research projects under way to find solutions for these issues. One promising approach is an already patented project,[10] in which a self-structuring, sustainable coating material made of wax and polysaccharide derivatives forms a superhydrophobic paper surface »**fig. 7**.[19]

Interaction of paper with humidity

Even water in a gaseous state can have a negative impact on building with paper, for example, if mould forms in an interior due to excessive humidity. To quantify the resistance of the paper material to water vapour, the so-called water vapour permeability of the paper material is characterised according to DIN 53122-1.[20] Here, the pore volume of the paper plays a decisive role. In this respect, its property can be influenced by paper-intrinsic factors, such as grammage or freeness of the pulp, but also by paper-extrinsic factors, such as a coating or sizing that changes the material structure and reduces the pore volume.

Humidity can also weaken the dimensional stability and mechanical properties of the material. If the air humidity changes, the equilibrium moisture content of the hygroscopic paper material is greatly altered. The fibres swell, and deformations occur. In addition, its mechanical properties deteriorate the wetter the paper becomes. Paper's susceptibility to moisture can be influenced by special drying processes during production, chemical additives, fibre shape or the amount of fill.

7 Extremely water repellent, superhydrophobic paper surface achieved with a wax and polysaccharide derivative coating.

Protective mechanisms against microorganisms

If cellulose comes into contact with water, not only does its stability suffer but this also creates an ideal breeding ground for microorganisms. Since cellulose is the very basis of all plants, various organisms can break down this basic substance both aerobically and anaerobically. The biological degradation of cellulose, in turn, reduces its mechanical stability.

Biocides can prevent this process by destroying, deterring, rendering harmless, preventing the action of or otherwise controlling the harmful organisms "by means other than mere physical-mechanical action". Their definition, authorisation and use are regulated by the European Biocidal Products Regulation.[22] The active substances are divided into four main groups concerning resistance risks and their effect: disinfectants, preservatives, pest control and others (taxidermist and antifouling agents). All information on active biocidal substances and authorised products is freely available.[23] Coating agents, fibre preservatives, slimicides and wood preservatives are suitable for protecting papers from harmful organisms.[24] In addition, research is conducted into the antimicrobial activity of cellulose and papers, for example, through secondary plant substances or naturally occurring barrier polymers, to be able to dispense with biocides in the future.

The flammability of paper

Like wood, paper (cellulose) is an organic, combustible (even flammable) material. If such materials are to be used for building purposes, fire protection precautions are required to protect them from destruction by fire.

Reaction to the fire behaviour of paper

Fire is generally defined as the formation of flames when organic substances and metals burn and give off light and heat. For a fire to start, three conditions and the correct mixture ratio of them must be met: sufficient fuel (combustible), sufficient oxygen (or another oxidant) and a minimum temperature (the so-called ignition point) **»fig. 8**. If one of the three prerequisites is removed from the burning process, the fire goes out.[25] This can happen if either the fuel or the oxygen component are consumed in the course of the fire or if sufficient cooling occurs, for example, by introducing extinguishing water.[26]

Additives for the fire protection of paper

Flame retardants protect paper from burning by improving its resistance to fire. They interrupt the combustion process in one or more complex stages. The aim here can be to prevent ignition; reduce the rate of fire; or influence the combustion mechanism in such a way that the combustion process is disturbed and, ideally, stopped. This disturbance of the combustion process can be triggered by a physical, chemical or a combined mechanism, depending on the type of flame retardant. The main flame retardant mechanisms are:[27]

- Cooling the substrate to a temperature below the necessary decomposition temperature, for example, by an endothermic reaction.
- Producing inert gases that lower the oxygen concentration at the surface of the substrate and thus smother the flame.
- Forming a protective coating that acts as a heat shield, which impedes oxygen diffusion in the decomposition area and prevents the diffusion of volatile combustible gases to the surface.

Furthermore, a distinction is made between flame retardants according to their nature or mode of action: halogen-, phosphorus-, nitrogen- and silicate-based flame retardants, as well as mineral flame retardants and so-called intumescent systems » **fig. 9**.

Halogen-based flame retardants are the most common: they are inexpensive, very effective, sufficiently available and proven.[28] Nowadays, however, many of these agents are banned or only permitted in minimal quantities as, on the one hand, they

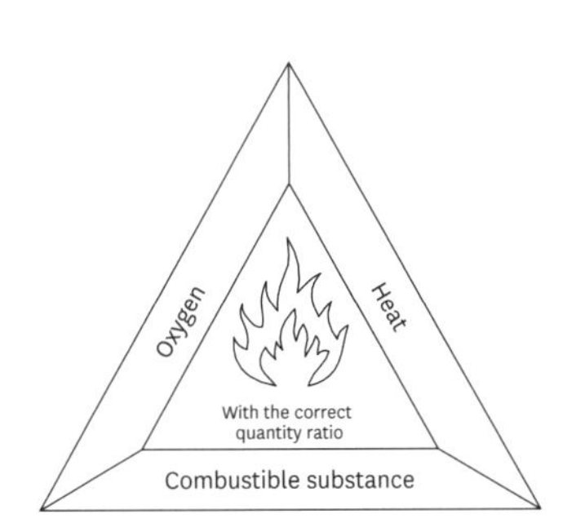

8 The combustion triangle names the conditions necessary for a fire to start and be maintained: heat, oxygen and fuel.

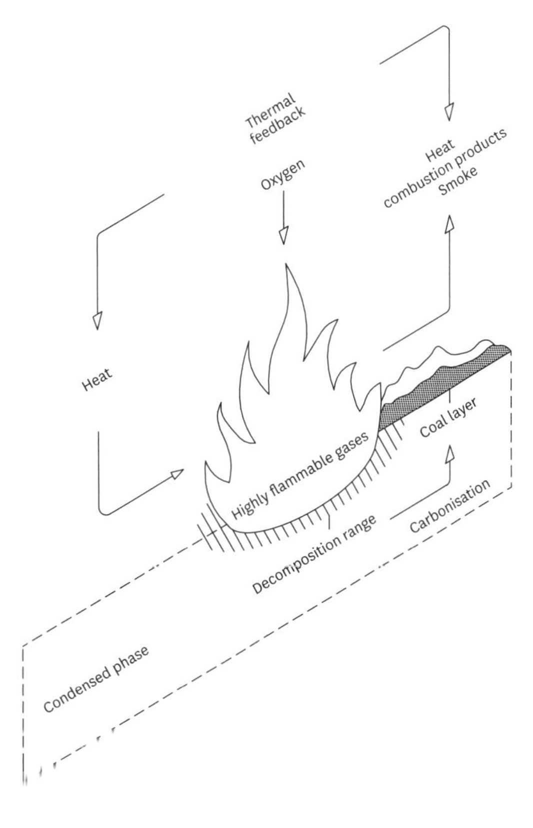

9 Mode of action of intumescent flame retardants: formation of a stable carbon layer that protects the material from the flames.

are harmful to the environment and, on the other hand, they release poisonous and acid gases when burned and can thus have a toxic effect on humans.[29]

An increasingly common alternative is phosphorus-based flame retardants. Their high versatility makes them suitable for many different applications. By forming phosphoric acid, they act as dehydrating agents and thus promote the process of charring. The char layer protects the remaining material from further decomposition.[30]

Phosphoric flame retardants form excellent synergies with nitrogen-containing flame retardants such as melamine and its derivatives. Inert gases are formed, which reduce the oxygen concentration and thus smother the fire. Combining both classes of flame retardants results in so-called intumescent systems (from Latin intumescencia – swelling, i.e. an increase in volume, for example, through foaming). Increased char production and the simultaneous production of inert gases create carbon foam in these systems. The foam acts not only as a heat shield against the fire but also as a diffusion barrier for oxygen and flammable gases. This removes the basic components of a fire, and the fire goes out. Ammonium polyphosphate (APP) is the best-known intumescent system.[31]

Mineral flame retardants are inorganic fillers that are introduced into the materials. The bound water, which is released during combustion, cools the fire as it evaporates and extinguishes it.

Silicate flame retardants work best in combination with other flame retardants. Although they can be used alone, they are not very effective compared with the other systems. The burning of silicate flame retardants forms a protective coating that serves as a diffusion and temperature barrier.

REFERENCES

1 *DIN 6730:2017-09: Paper, Board and Pulps – Terms*, Berlin: Beuth, 2017.
2 *DIN 6735:2010-02: Paper, Board and Pulps – Survey of Terms and Definitions*, Berlin: Beuth, 2010.
3 Dieter Klemm, Brigitte Heublein, Hans-Peter Fink, Andreas Bohn, "Cellulose: Fascinating Biopolymer and Sustainable Raw Material", in: *Angewandte Chemie* 44(22), 2005, pp. 3358–3393.
4 Peter Zugenmaier, *Crystalline Cellulose and Cellulose Derivatives: Characterization and Structures*, Berlin/Heidelberg: Springer Verlag, 2008.
5 Dieter Klemm, Bertrand Philipp, Thomas Heinze, Ute Heinze, W. Wagenknecht, *Comprehensive Cellulose Chemistry; Volume 1: Fundamentals and Analytical Methods*, Weinheim: Wiley-VCH Verlag, 1998.
6 Verband deutscher Papierfabriken e. V. – VDP, Kennzahlen Zellstoff- und Papierfabriken in Deutschland 2020/2019, Bonn, 2020, https://www.vdp-online.de/industrie/statistik, accessed 19 March 2021.
7 Verband deutscher Papierfabriken e. V. – VDP, Papier Kompass, Bonn, 2020, https://www.vdp-online.de/fileadmin/0002-VDP/07_Dateien/7_Publikationen/Kompass_de.pdf, accessed 19 March 2021.
8 *Papier 2022 – Ein Leistungsbericht*. Verband Die Papierindustrie e. V., Bonn, 4 May 2022.
9 Verband deutscher Papierfabriken e. V. – VDP, Gregor Andreas Geiger, *Papiermachen*, Bonn, 2015, https://www.vdp-online.de/filead-[illegible] Publikationen/Papiermachen.pdf, accessed 19 March 2021.
10 Holger Burkert, "Die Aufrollung", in: *Papier und Technik*, 2017, https://www.papierundtechnik.de/papiertechnik/die-aufrollung-2/, accessed 25 October 2022.
11 *DIN 54514:2008-08: Testing of Paper and Board – Determination of Initial Wet Web Strength by Tensile Test*, Berlin: Beuth, 2008.
12 Monica Ek, Göran Gellerstedt, Gunnar Henriksson (eds.), "Paper Products: Physics and Technology", in: *Pulp and Paper Chemistry and Technology*, Volume 4, Berlin: DeGruyter, 2009.
13 Jürgen Blechschmidt (ed.), *Taschenbuch der Papiertechnik*, 2nd edition, Munich: Carl Hanser Fachbuchverlag, 2013.
14 *DIN EN ISO 535:2014-06: Paper and Board – Determination of Water Absorptiveness — Cobb Method (ISO 535:2014)*, Berlin: Beuth, 2014.
15 Wilhelm Barthlott, Zdenek Cerman, Anne Kathrin Stosch, "The Lotus Effect: Selbstreinigende Oberflächen und ihre Übertragung in die Technik", in: *Biologie in unserer Zeit*, 34(5), 2004, pp. 290–296.
16 Hongta Yang, Yulin Deng, "Preparation and Physical Properties of Superhydrophobic Papers", in: *Journal of Colloid and Interface Science*, 2008, 325(2), pp. 588–593.
17 Andreas Geissler, Florian Loyal, Markus Biesalski, Kai Zhang, "Thermo-Responsive Superhydrophobic Paper Using Nanostructured Cellulose Stearoylester", in: *Cellulose*, 2014, 21(1), pp. 357–366.
18 Andreas Geissler, Markus Biesalski, Regenerierbare superhydrophobe Beschichtungen, Patent WO 2018/193094 AI, 2018.
19 Cynthia Cordt, Andreas Geissler, Markus Biesalski, "Regenerative Superhydrophobic Paper Coatings by In Situ Formation of Waxy Nanostructures", in: *Advanced Materials Interfaces*, 8(2), January 2021, https://doi.org/10.1002/admi.202001265
20 *DIN 53122-1:2001-08: Testing of Plastics and Elastomer Films, Paper, Board and Other Sheet Materials – Determination of Water Vapour Transmission – Part 1: Gravimetric Method*, Berlin: Beuth, 2001.
21 Monica Ek, Göran Gellerstedt, Gunnar Henriksson (eds.), "Paper Products: Physics and Technology", op. cit.
22 Regulation (EU) No 528/2012 of the European Parliament and of the Council of 22 May 2012 concerning the making available on the market and use of biocidal products. Text with EEA relevance, here Article 3.
23 Federal Institute for Occupational Safety and Health (n.d.), Helpdesk: Reach – CLP – Biocide: Approved Active Substances, https://www.reach-clp-biozid-helpdesk.de/DE/Biozide/Wirkstoffe/Genehmigte-Wirkstoffe/Genehmigte-Wirkstoffe-0.htm, accessed 23 June 2020. European Chemicals Agency (n.d.), Information on chemicals – Suppliers of active substances, https://echa.europa.eu/de/information-on-chemicals/active-substance-suppliers, accessed 23 June 2020. Fungicide Resistance Action Committee – Publications, https://www.frac.info/publications/downloads, accessed [illegible]
24 Commission Implementing Regulation (EU) 2018/1602 of 11 October 2018 amending Annex I to Council Regulation (EEC) No. 2658/87 on the tariff and statistical nomenclature and on the Common Customs Tariff, Part 2, Section X, Chapters 47 and 48.
25 D. K. Shen, Sai Gu, "The Mechanism for Thermal Decomposition of Cellulose and its Main Products", in: *Bioresource Technology*, 100(24), 2009, pp. 6496–6504.
26 Arvind Atreya, "Ignition of fires", in: *Philosophical Transactions of the Royal Society of London. Series A: Mathematical, Physical and Engineering Sciences*, 356(1748), 1998, pp. 2787–2813.
27 Constantine Papaspyrides, Pantelis Kiliaris (eds.), *Polymer Green Flame Retardants*, Amsterdam: Elsevier, 2014.
28 Edward D. Weil, Sergei V. Levchik, *Flame Retardants for Plastics and Textiles: Practical Applications*, 2nd edition, Munich: Carl Hanser Verlag, 2015.
29 Susan Shaw, "Halogenated Flame Retardants: Do the Fire Safety Benefits Justify the Risks?", in: *Reviews on Environmental Health* (25)4, 2010, pp. 261–306.
30 Ike van der Veen, Jacob de Boer, "Phosphorus Flame Retardants: Properties, Production, Environmental Occurrence, Toxicity and Analysis", in: *Chemosphere* 88(10), 2012, pp. 1119–1153.
31 Sergei V. Levchik, Giovanni Camino, Luigi Costa, G. Levchik, "Mechanism of Action of Phosphorus-Based Flame Retardants in Nylon 6. I. Ammonium Polyphosphate", in: *Fire* [illegible]

3 SEMI-FINISHED PRODUCTS AND COMPONENTS

This chapter presents the different types of paper, semi-finished products and components suitable for use as building materials. A distinction is made between flat and rod-shaped (linear) semi-finished products and components. In addition, specific components will be presented that have been specially developed for construction use.

Material parameters must be defined to enable the use of these materials in structural design. The following chapter describes the essential material properties with the typical values for building materials to get a first impression of the different types of paper, semi-finished products and components.

The production process of the materials is decisive for their basic properties. Therefore, the various manufacturing processes of the respective semi-finished products and components are also briefly described.

Paper, cardboard and paperboard

As described in chapter 2, paper is available as cut-to-size sheets or as rolled goods. The papermaking machine determines the maximum width, and the material properties will already be adjusted to the intended function during production.

Papers and cardboards are flat semi-finished products – they differ only insignificantly from each other, as their raw materials and production processes are almost identical. DIN 6730 differentiates between paper and board according to the grammage or weight per unit area: from 225g/m² upwards, the standard defines the semi-finished paper product as a board, below that as paper.[1]

The strength of a paper can be increased and its water-repellent properties improved by using synthetic fibres instead of conventional wood fibres in the manufacturing process or by mixing them in the pulp proportionally. Typical applications are ID cards, driving licences and waterproof cards.[2] In the construction industry, where robust and waterproof papers are virtually a prerequisite, the admixture of synthetic fibres could be of great significance.

For some applications, the surface quality of the paper or board coming out of the papermaking machine is insufficient. Additional processing steps such as satinating or coating the paper can enhance the surface quality. Some papers are embossed to create a texture in the paper surface or coated with plastics or varnishes to make the material waterproof. Paper components with embossed surfaces are easier to combine with other materials. For example, mineral materials such as clay or plaster adhere much better to an embossed surface.

Lamination is the process of joining paper products with different materials – such as other paper grades, plastics or metal foils – to improve its properties. Several layers of paper of the same or different quality can be glued or laminated together to obtain different surface qualities on each side. The combination of paper, plastics and metal foils is familiar from household items that protect food from moisture, light exposure and odours. One typical example is liquid packaging board, whose paper is coated with plastics such as polyethylene (PE) on both sides. The packaging remains watertight but also prevents moisture from penetrating from the outside. Thin aluminium foils laminated to paper or a paper-plastic composite, on the other hand, prevent the penetration of light and oxygen.

Such optimised paper grades are ideal for the outer weather protection layer of a building and vapour barrier layers within a building component; they are also suitable for interior surfaces, for example, in wet areas. Furthermore, such optimisation can prevent microorganisms from decomposing the paper.

Nowadays, the lion's share of paper and board goes into paper printing and the packaging industry. Due to the numerous possibilities for modifying papers with fillers and additives, several thousand paper grades have been developed for specific uses. One basic property of paper that must be taken into account when selecting grade and location of use is anisotropy. The modulus of elasticity and bursting strength strongly depend on the running direction of the papermaking machine, which is of great importance for further processing.

Two corrugated base papers »**chapter 8, figs. 20, 21, pp. 182–183**, exemplify the choice of material: the liner forms the top layer of a corrugated board, which must be able to absorb tensile and compressive forces. Kraft paper is suitable for this purpose, i.e. paper made from a high-quality kraft pulp (a pulp with highly tear- and burst-resistant fibres). By increasing the grammage of the paper, its bending stiffness and resistance to puncture can be increased. The primary task of the corrugated layer in corrugated board is to keep these cover layers at a distance from each other. The strength requirements are less stringent here, which is why poorer paper qualities also suffice. The air pockets formed between the corrugated layer and the liner give the corrugated board very good thermal insulation properties. In addition to the fibre material, chemical additives can also significantly influence properties such as wet and dry strength. In chapter 8, the material properties of the two corrugated base papers (top liner and fluting) are listed as examples to provide an initial calculation basis for the design and dimensioning of a paper construction.

Analogous to corrugated board, paperboard is usually made up of several layers as well. High bending stiffness, in particular, is a typical requirement for folding box boards, and it is just as essential for structural engineering applications. Papermaking machines for paperboard often have several headboxes (i.e. several introduction points for different types of stock), each forming a single paper layer. It is, therefore, possible to combine different paper qualities into a paperboard by so-called couching, i.e. joining in the wet state, without having to use glue because, in the wet state, the paper fibres bond via hydrogen bonds »**section "Wet end", p. 26**. The multi-layer structure can be used to adjust the surface properties for a specific purpose. Accordingly, the front and back must not be confused or interchanged.

Cardboard and paperboard can also show a more or less pronounced anisotropy. As an example, chapter 8 lists the properties of a paperboard material »**chapter 8, fig. 22, p. 184** with good formability. This property is particularly beneficial for the production of special components such as connecting elements or design elements such as façade panels. Cardboard also comprises several thicker layers. The cardboard material listed as an example in »**chapter 8, fig. [illegible], p. [illegible]** consists mostly of wood pulp. It

1 Types of corrugated board

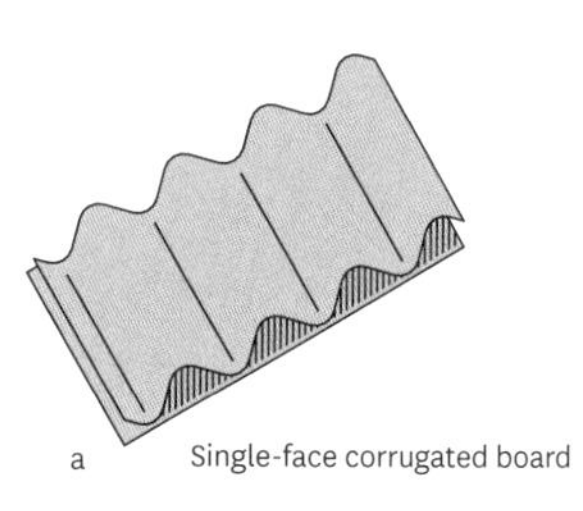

a Single-face corrugated board

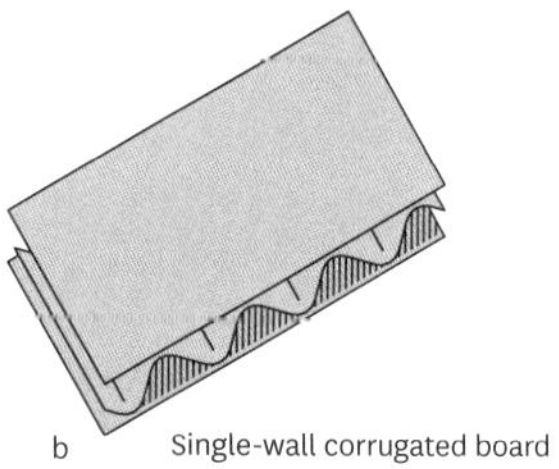

b Single-wall corrugated board

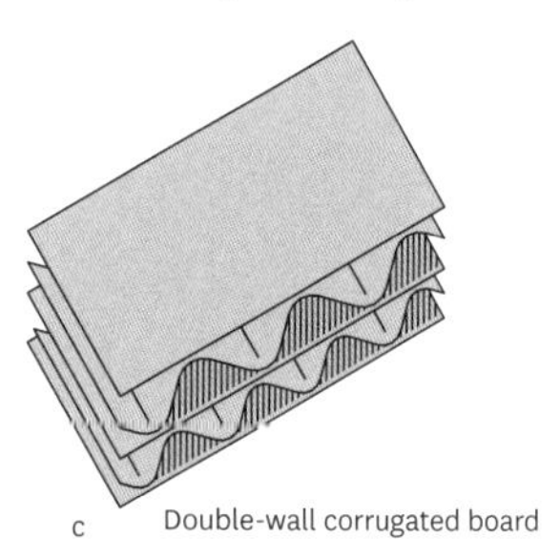

c Double-wall corrugated board

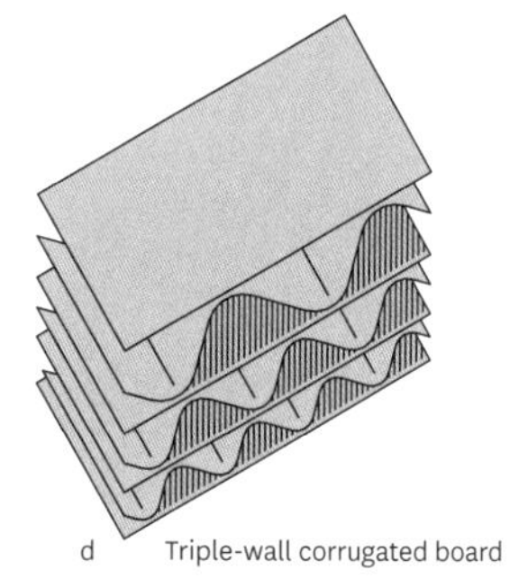

d Triple-wall corrugated board

is, therefore, voluminous and resilient and is already used in construction as impact sound insulation. The voluminous cardboards are also much more suitable for solid components than paperboard or paper are, as the higher volume means that fewer layers have to be joined. Another advantage is that it is still a mono-material, meaning there is less mixing of materials. It is, therefore, more ecological and easier to recycle. In addition, there are fewer potential sources of error or weak points in the construction, which might occur with glueing due to inhomogeneous glue application.

All the materials mentioned can be finished or refined in-line, i.e. directly on the papermaking machine, or off-line in downstream processes. Further processing is usually carried out with the materials as rolled goods.

Corrugated board

Corrugated board is one of the most widely used packaging materials in Europe. This is mainly due to its high stability combined with low density and low manufacturing costs. These properties are also relevant for structural applications. Because corrugated board has been optimised for transport and packaging purposes, it features good insulating properties in addition to high stability.

A special characteristic of corrugated board is its structure-related anisotropy. Corrugated board, therefore, shows different properties in different orientations. Parallel to the direction of the flutes, it is very stable against bending forces. However, if loaded perpendicularly to the direction of the flutes, corrugated board can easily buckle. In this respect, too, the strategies of the packaging and transport industries can serve as a model for building-technology applications. By adapting the geometry, e.g. folding, the risk of buckling can be reduced and the entire system's stability can be increased.

DIN 6735 defines the structure of corrugated board as "cardboard consisting of one or more layers of corrugated paper glued to one layer or between several layers of other paper or cardboard".[3] A distinction is made between single-face, single-wall, double-wall and triple-wall corrugated board **»fig. 1**.

Multi-layer corrugated boards often form the core in multi-wall corrugated boards; described in more detail in the **»section "Sandwich structures", pp. 40–41**. For corrugated board with multiple layers, not all layers have to have the same flute shape and flute type, as each performs a different function. For packaging, coarse or medium flutes are often combined with fine flutes: the flexible coarse flute protects the stored or transported goods, while the fine flute gives the corrugated board greater rigidity and protects it from external mechanical stresses. If this principle is transferred to the building industry, the medium waves can take over the heat-insulating properties and the fine waves the mechanical or structural properties of the material.

The flutes can have different shapes and dimensions. The flute can have a U-shape, V-shape or sine-shape, whereby the sine-shape is a combination of the two others. It combines the advantages of the U- and the V-shapes **»fig. 2**.

Composition and production of corrugated board

The properties of corrugated board are influenced not only by the number of layers and the combination of flute types and shapes but also by the chosen combination of materials. The papers used in corrugated board are divided into liner papers and fluting papers based on their different functions **»fig. 3**. The fluting paper is protected inside

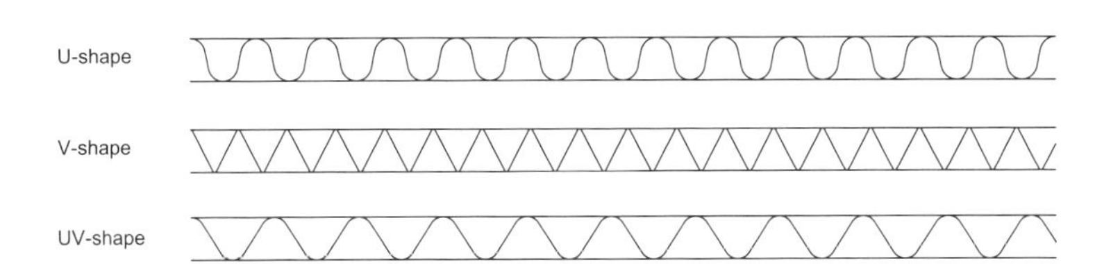

2 Flute shapes of corrugated board.

the corrugated board structure by the outer layers and thus very rarely needs to have water-repellent or optical properties. The cover paper, on the other hand, is often printed and coated against moisture.

Fluting papers are classified according to semi-chemical pulp and corrugating medium. Semi-chemical pulp papers consist mainly of wood fibres that are broken down semi-chemically. They must not contain more than 35% recovered paper. Corrugating medium, on the other hand, consists exclusively of recycled fibres. Liner papers are differentiated according to the types of kraft liner, test liner and Schrenz liner. If particular importance is attached to high stability and strength, kraft liner is used because it consists mainly of kraft pulp and features high strength. Test liner is an alternative for liner papers that have to meet lower quality requirements. The material is more cost-effective and consists entirely of recycled waste-paper fibres. Schrenz paper consists of unsorted waste-paper fibres and thus has the lowest load-bearing capacity.

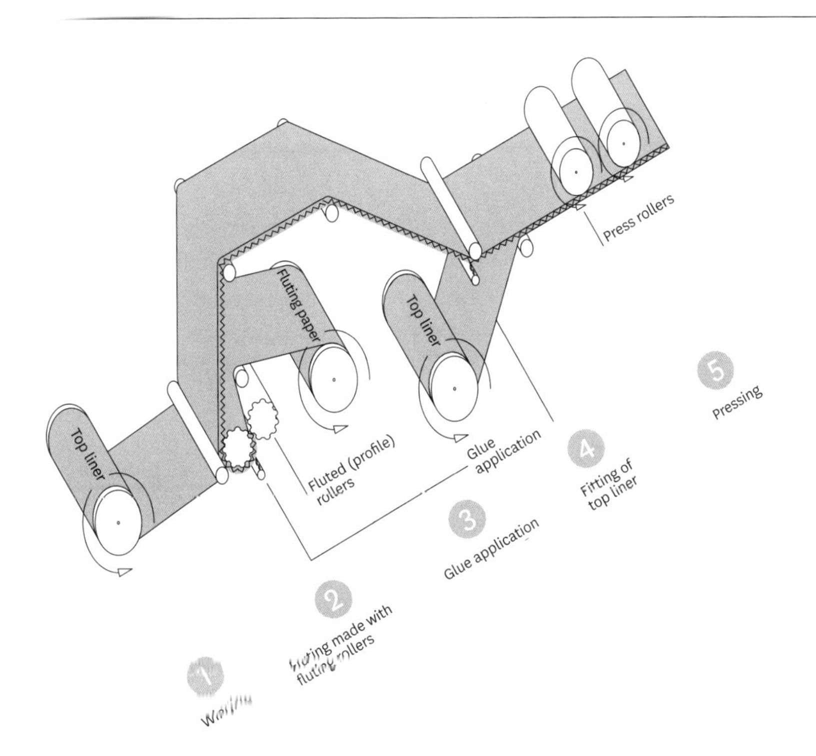

3 Schematic diagram of corrugated board production (single-wall corrugated board).

4 Structure of sandwich panels, a. solid core, b. honeycomb core.

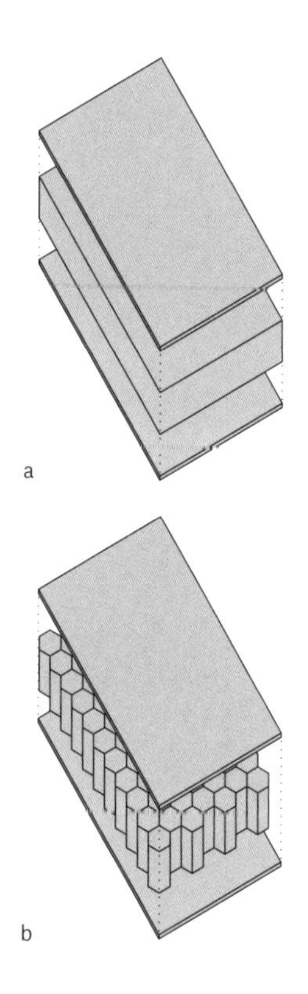

In addition to the wide range of paper types, there are many adhesives available for production, roughly differentiated into waterproof and water-soluble. Glues based on potato, maise and wheat starch are the most common. They can also be combined and wet-strengthened, i.e. made water-resistant, by using additives.[4] After applying the starch-based glue to the paper layers, the liquid parts of the glue dry out and a stable bond is formed. Between 16 to 40g/m² of glue is applied during glueing, depending on the chosen paper combination. After drying, the glue has a surface-related mass of approx. 4 to 10g/m² because a large part of the liquid evaporates from the glue.[5]

Properties of corrugated board

As previously described, the properties of corrugated board depend on the materials used for the cover and intermediate liners and fluting papers as well as the type of flute. Accordingly, designers have any number of possibilities to arrange the corrugated boards in such a way that the required demands are met. DIN 55468-1 defines mechanical requirements for corrugated board such as burst strength; puncturing energy, i.e. stability against a sudden point load from a sharp object; and edge crush resistance.[6] The mechanical properties relate to the requirements of packaging – the most common use of corrugated board to date. These properties provide reference values for the building-technical or structural suitability of the different types of corrugated board.

The different types of paper described in the previous chapter are suitable for both the liners and the fluting. They influence the properties of the corrugated board and enable numerous material combinations and corrugated board structures. » **Chapter 8, fig. 24, p. 186,** provides a table with structural data for a single-face corrugated board.

Sandwich structures

For higher and direction-independent bending stiffness, sandwich structures **» fig. 4** are the better choice; they are far sturdier than corrugated boards and are no more susceptible to bending loads in one direction than another. Honeycomb boards belong to the sandwich or composite boards; they typically consist of two face sheets and a core. Unlike corrugated boards, honeycomb boards also have direction-independent bending properties.

5 Honeycomb core of a sandwich structure with hexagonally shaped cells.

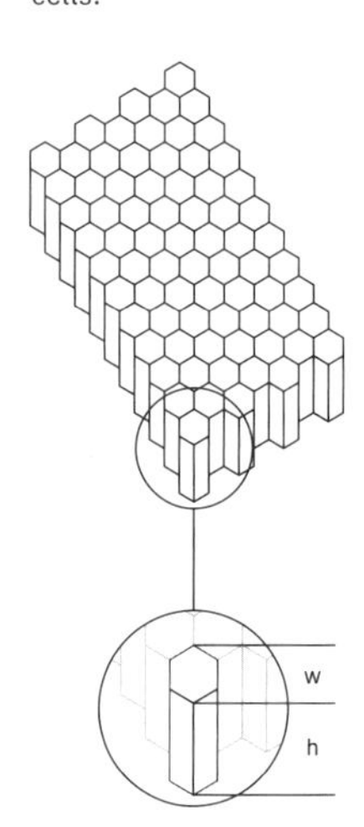

The face sheets form the load-bearing structure, which usually requires stiff and robust material. They protect the core from external forces and absorb possible tensile loads. The core, located between the two face sheets, supports these sheets, regulates the distance between them, and creates the shear strength of the composite structure.

Sandwich panels are often used when a combination of high stiffness and low weight is required. The ratio resulting from the two values is referred to as the strength-to-weight ratio. To achieve an optimum strength-to-weight ratio, the core must create the greatest possible distance between the face sheets because the larger the distance between the face sheets, the higher the sandwich panel's bending stiffness. At the same time, the greater the distance between the two face sheets, the higher the core's weight share in the total weight of the sandwich panel, which ultimately changes the mechanical stress. Therefore, the materials and dimensions of the face sheets and the core need to be carefully calculated and selected.

The mechanical properties of the honeycomb core result from the choice of material, the material thickness, the cell height and the cell width – “h” and “w”, respectively, in »fig. 5. The design of a honeycomb panel always requires a compromise between stiffness and weight: the stiffer the panel, the higher its weight because stability requires either higher material thickness or reduced cell size. As an alternative to a honeycomb core made of hexagonal cells, corrugated cardboard in the form of multi-wall corrugated boards can also be used as core material. »Fig. 6 shows the production of such multi-wall corrugated boards.

In the building industry, one important factor is the insulating effect of façade elements. The advantage here is the low thermal conductivity of paper as well as the air trapped in the material, both of which contribute to its effectiveness as a thermal insulation material. Sandwich structures are also well suited as sound- and heat-insulating floor coverings in living spaces when pressure resistance in the vertical direction is advantageous. However, to avoid damage to the surface, the material of the top sheet must be able to withstand point loads well – for example, from stiletto heels.

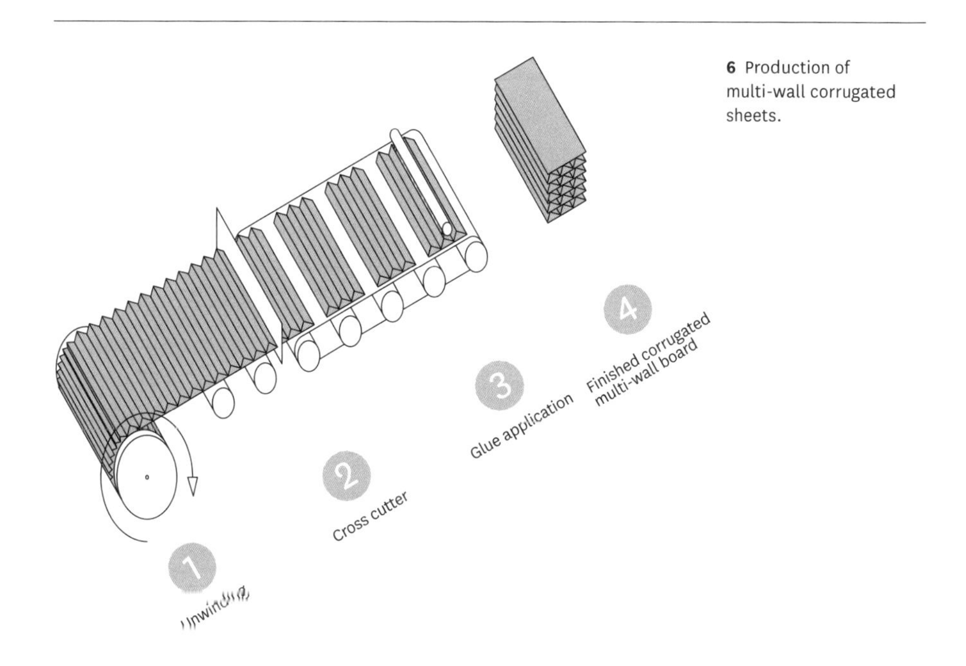

6 Production of multi-wall corrugated sheets.

Tubes and cores

Industrially manufactured, rod-shaped components made of paper are created by glueing together several layers of paper, paperboard or cardboard. Common types are wound tubes or laminated profiles » **section "Profiles", pp. 45–46**.

Production of tubes and cores

A distinction is made between three manufacturing processes: parallel or convolute wound, perpendicular or vertical wound and spiral wound cores **» fig. 7**. The last-named dominate the market due to their continuous and thus more cost-effective winding technique. The predominantly longitudinally oriented fibres in parallel-wound cores offer a mechanical advantage when used as beams or supports – although this is only slightly pronounced with the conventionally used, non-highly oriented papers. Compared with spiral wound cores, parallel wound cores are limited regarding their dimensions and the combination of different papers.

7 Basic methods of winding around a mandrel. The running direction of the paper is marked with an arrow.

Winding type

Parallel or convolute winding

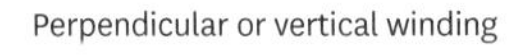

Spiral winding

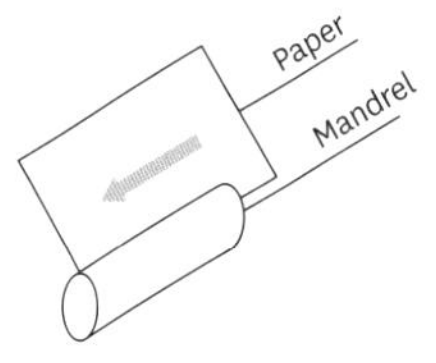

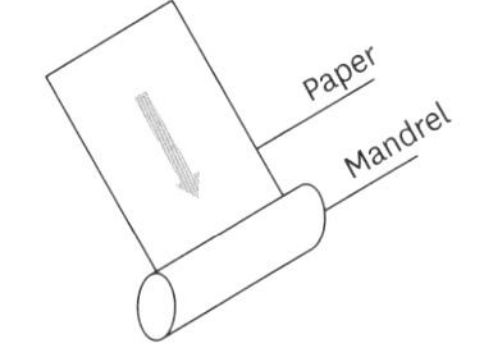

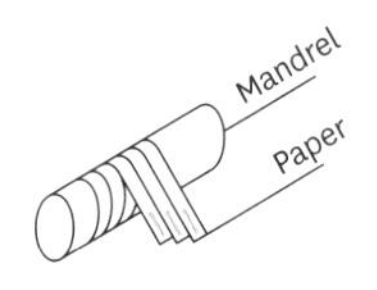

8 Winding methods for papers and glues suitable for building construction.

Winding type	**Parallel or convolute winding**	**Perpendicular or vertical winding**	**Spiral winding**
Paper	· core winding paper or paperboard · soda paper or paperboard · kraft liner	· core winding paper or paperboard · soda paper or paperboard · kraft liner	· predominantly fourdrinier board (coreboard) with a thickness of 0.2mm to 1mm and a grammage of 150g/m² to 650g/m² · soda paper · kraft liner · test liner
Adhesives	· casein glue · starch glue · silicate glue · dispersion adhesives · bone glues	· casein glue · starch glue · silicate glue · dispersion adhesives · bone glues	· polyvinyl acetate (PVAc) adhesive; limited water resistance (can be increased by adding cross-linking agent) · polyvinyl alcohol (PVA) adhesive; higher initial tack compared with starch and cellulose adhesives (tack, and also water resistance, can be increased by adding cross-linking agents) · dextrin and starch glues are water-soluble · silicate- and casein-based glues (casein is water-soluble)

Depending on the intended use; the requirements of the products; and, of course, the manufacturing processes, different adhesives and paper grades with different lengths of fibres and different paper thicknesses are used »**fig. 8**.

Properties of tubes and cores

The different winding processes limit the dimensions that can be produced, influence the orientation of the usually slightly anisotropic material and affect the geometric and structural requirements of the component »**fig. 9**. While spiral winding allows for endless paper tubes »**fig. 12**, the lengths offered by parallel and vertical winding are limited by the dimension of the starting material or mandrel. Theoretically, a mandrel could be

9 Comparison of the three manufacturing processes regarding properties, load types and areas of application.

Winding type	**Parallel or convolute winding**	**Perpendicular or vertical winding**	**Spiral winding**
Dimensional limitations	· length of the component is limited by mandrel length · thickness and diameter of the component are limited by the width of the paper sheet	· length of the component is limited by the width of the paper sheet	· easy adjustment of the wall thickness · limitation of the thickness of the core (currently 1 to 30mm) by the machine-specific possible number of processable paper sheets (current maximum: 36, see **fig. 12**) · production of long lengths possible through endless winding (currently up to 11m), limitation mostly due to premises size
Orientation of the machine direction of the paper	· oriented in the longitudinal axis of the core Paper Mandrel	· oriented in the circumferential direction of the core Paper Mandrel	· oriented at different angles in the circumferential direction of the core Mandrel Paper
Core properties	· poor concentricity properties due to the characteristic seam joint on the inside and outside of the core (grinding possible) · high breaking resistance in the longitudinal axis	· poor concentricity properties due to the characteristic seam joint on the inside and outside of the core (grinding possible)	· very good concentricity properties · different materials per layer can be used · lower breaking resistance in the longitudinal axis, as the joint constitutes a defect
Preferred load type	· bending load of the core · compressive load on the cross-sectional area	· external and radial pressure	· depends on the winding angles of the paper sheet
Typical use	· as roll cores, mainly for flexible fabrics (textile industry)	· fireworks · cartridge cases · cardboard barrels (fibre drum) · thick-walled tubes (> 20mm wall thickness)	· as roll cores, e.g. in the paper and plastics industry · packaging material, e.g. as a drum or can

10 Spiral-wound tube: They are often used as cores or as packaging.

11 Spiral-wound tube with another paper material.

made 20m long; however, this would not be economical. In practice, the usual length of parallel winding is about 1 to 3m, and approx. 0.05 to 1m with vertical winding. The world's largest machine currently produces paper sheets 10m wide, which would allow for components 10m long made with vertical winding.

In current industrial manufacturing, only spiral wound tubes are produced in sufficient length and wall thickness to make them suitable for use as construction material. However, since the circumferential seam joint is the weak point of the material, spiral wound tubes have a lower breaking resistance in the longitudinal axis compared with tubes made with the other winding methods. In this respect, parallel wound tubes feature particularly good properties. But even though parallel wound tubes have the greatest potential for use as beams or supports, their availability on the market in the quantities and dimensions needed remains insufficient because there is too little demand.

» Figs. 10, 11 show spiral-wound cores. By using different base materials, the properties and surface textures of the tubes can be changed, for example, printed or coated papers can be used as surface material. In architecture, tubes became known mainly through Shigeru Ban and his projects, such as the Japanese Pavilion **» pp. 132–133**. He makes use of the fact that tubes can be produced in long lengths through the winding process. Usually, the tubes are impregnated with water-repellent materials before installation. Theoretically, however, it would also be possible to apply a water-repellent-coated paper as the top layer during the winding process. In this way, the winding processes can be used to combine different types of paper according to requirements and needs in order to create certain properties, such as special surface textures.

Profiles

In addition to wound tubes, laminated profiles in I-, L- or U-shape are also among the industrially available rod-shaped semi-finished products. Currently, they are primarily used as transport protection.

For this application, profiles have already been optimised in terms of stability and durability. The experience from this field benefits the structural engineering application and allows technical dimensioning or at least well-grounded estimations.

Next to tubes, profiles are rod-shaped elements that can be used particularly well as load-bearing components due to their high load-bearing capacity. For structural use, profiles have the significant advantage of flat surfaces compared with the curvatures of tubes, which makes it much easier to connect them to flat building components using conventional fasteners such as screws.

If, for example, profiles are joined together to form a square cross-section to serve as a beam, and sheets of corrugated or honeycomb board are attached to it as insulation and building envelope, no elaborate fasteners are necessary.

Manufacture of cardboard profiles

Cardboard profiles are manufactured by wetting several cardboard strips with adhesive, stacking them on top of each other and finally enveloping them with an enclosing cover layer of paper. Subsequent pressing over two rollers improves the bonding and dimensional accuracy of the laminate. To produce L- or U-shaped profiles, additional profile rollers must be installed downstream before the adhesive cures. In particular to avoid damaging the cardboard, forming is done in several steps. A drying zone follows the profiling process. The continuous profiles can then be cut to order using a cutting unit » **fig. 13**.

Properties of cardboard profiles

The columns or posts commonly used in construction are components loaded with longitudinal force or with longitudinal force and bending. The performance capabilities of typical laminated cardboard cross-sections are listed in the two tables in » **figs. 14, 15**.

Both tables list possible beam constructions with corresponding loads and qualitative assessment. In addition, implementation examples for the material paper are given for the different support types.

Similar to core winding, the paper quality and the type of adhesive can be adjusted for laminated profiles according to the desired end product. With conventional profiles, the paper fibres are always oriented in the longitudinal direction and cannot be adjusted. In the construction industry, this is an advantage for applications with tensile or compressive loads on the cross-section, as the load is applied according to the orientation of the paper fibres.

Natural fibre-reinforced sheets

In contrast to the industrially manufactured products mentioned above, research is still being conducted on papers reinforced with natural fibres. They consist of a paper fibre matrix and a natural fibre reinforcement (flax, jute, etc.) made with paste or starch as an adhesive agent. Reinforcing the paper with high-strength natural fibres is intended to improve its mechanical properties and qualify it for use in load-bearing components. Specifically, this is about natural fibres in the form of semi-finished textile products, such as unidirectional fibre mats or fabrics made of flax or jute. These products are also used for fibre-plastic composites, for example, and have good mechanical properties. Unidirectional natural fibre mats give the new composite material anisotropic – i.e. direction-dependent – material properties, which also characterise the fibre-plastic composite solutions.

Natural fibre-paper composite usually comprises several individual layers, each consisting of a combination of paper fibres and natural fibre mats, alternately layered on top of each other to form a laminate » **fig. 16**. The mechanical behaviour of the laminate is influenced by the chosen material and by its composition.

Materials

Starting materials have a significant influence on the properties of a composite material. It should be noted that when a single layer is subjected to tensile and compressive loads in the direction of the fibres, the reinforcing fibres bear a large part of the load. When a single layer is subjected to loads crosswise to the fibre orientation, the paper fibres have to absorb the loads. In the case of compressive loads, the paper fibre matrix should support the natural fibres/reinforced fibres and hold them in position.

The mechanical properties of a composite also depend on the fibre volume fraction of the reinforcing fibres. Generally speaking, the reinforcing fibres have significantly better mechanical properties than the paper fibre matrix – especially in the fibre direction. Thus, the mechanical properties of the composite improve according to the proportionate share of reinforcing fibres in it.

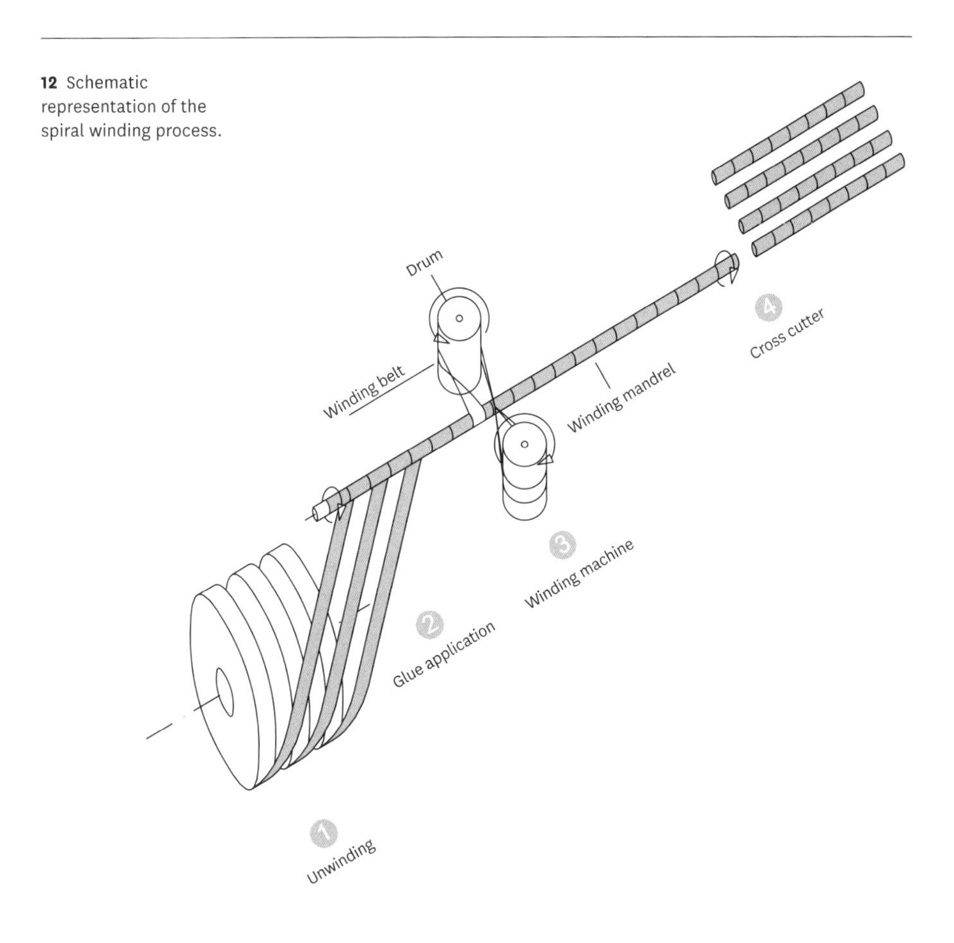

12 Schematic representation of the spiral winding process.

13 Industrially available profiled cardboard or paperboard.

Profile shape	**I-profile**	**L-profile**	**U-profile**
Width	b_{min} = 40mm b_{max} = 160mm		b_{min} = 8mm b_{max} = 200mm
Leg length		a_{min} = 25mm a_{max} = 150mm	a_{min} = 12mm a_{max} = 100mm
Thickness	t_{min} = 2mm t_{max} = 10mm	t_{min} = 2mm t_{max} = 10mm	t_{min} = 2mm t_{max} = 10mm
Length	l_{max} = 8000mm	l_{max} = 8000mm	l_{max} = 8000mm

14 Overview and assessment of possible beam constructions.

Support type	**Solid beam**		**Thin-walled closed beam**	
Cross-section	Round or square	Any rectangle (aspect ratio not equal to 1)	Round or square	Any rectangle (aspect ratio not equal to 1)
Schematic diagram				
Load type	Absorbs longitudinal forces	Absorbs longitudinal forces	Absorbs longitudinal forces	Absorbs longitudinal forces
Exemplary implementation in paper	Laminated or wound solid beam	Laminated solid beam	Wound beam or beam constructed from profiles	Wound beam or beam constructed from profiles
Assessment	· Entire cross-section absorbs longitudinal forces · Particularly the outer edge layers of the solid beam provide protection against buckling of long components · Material cannot be fully utilised	· Entire cross-section absorbs longitudinal forces · Buckling first occurs perpendicular to the smallest main beam axis · Material cannot be fully utilised	· Material only where it is needed · Thinness can lead to local bulging · Entire cross-section absorbs longitudinal forces · Particularly the outer edge layers of the beam provide protection against buckling of long components · Material can be fully utilised	· Entire cross-section absorbs longitudinal forces · Buckling first occurs perpendicular to the smallest beam axis · Material cannot be fully utilised

Laminate composition

The mechanical properties of a composite can also be influenced via the laminate composition (number of layers, orientation of the fibres and thickness of the layers). The composition and orientation of the individual layers can be chosen freely. However, some design rules must be observed, such as retaining a symmetrical and balanced laminate structure to avoid undesirable side effects like twisting of the laminate. One typical approach is to stack the individual laminate layers such that the fibres in each layer are orthogonally offset from those in the next layer to compensate for anisotropic material properties. The more layers and the more different fibre orientations, the more isotropic the laminate becomes.

15 Overview and assessment of possible beam constructions made of paper.

Support type	**Solid beam**	**Thin-walled beam**		**Trussed beam**	**Frame girder**
		without flange	**with flange**		
Schematic diagram					
Load type	Wall absorbs longitudinal forces and thrust in through-thickness direction → triaxial state of stress	Skin takes over longitudinal forces and intralaminar thrust in the plane → plane state of stress	Wall absorbs intralaminar thrust in the plane and flange absorbs longitudinal forces	Beams with longitudinal forces	Frame girder with longitudinal forces and bending moments
Exemplary implementation in paper	Laminated solid beam	Wound beam	Flanges with winding	L-profile truss	Cut-out cardboard
Sensitivity to torsion	No	No	No	Conditional	Yes
Assessment	· Interlaminar loading of the paper · Material cannot be fully utilised	· Shear and bending stiffness cannot be adjusted independently of each other · Material cannot be fully utilised · Prone to bulging	· Separate material-compatible design of belt and skin possible	· Separate design of tension and compression rods possible	· Thrust is generated by the bending stiffness of the frame girder · Shear and bending stiffness cannot be adjusted independently of each other · Material cannot be fully utilised

16 Exemplary composition of a laminate,
a. unidirectional single layer,
b. laminate made of several individual layers.

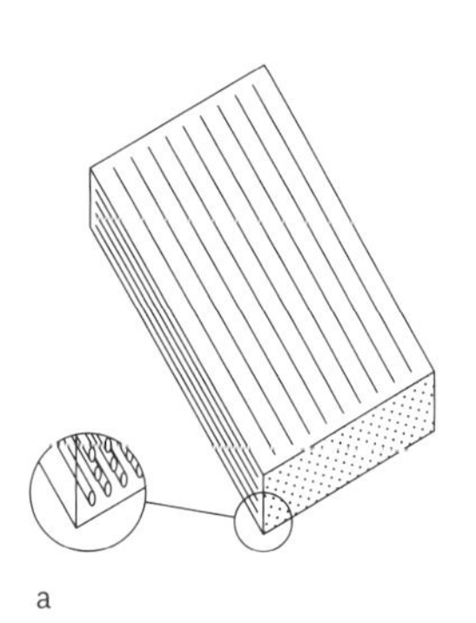

a

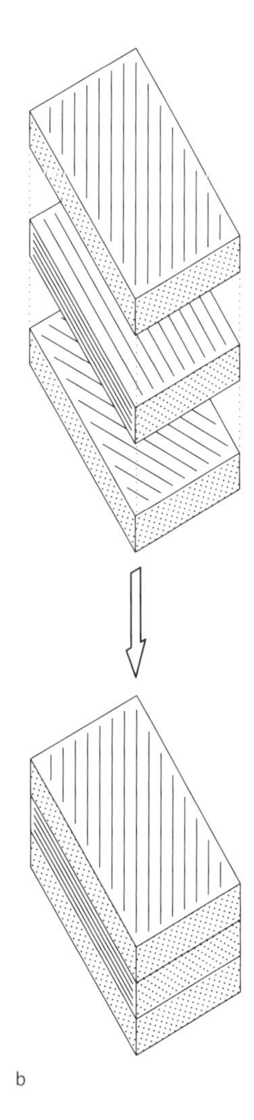

b

Production of natural fibre-reinforced papers

In natural fibre-reinforced paper production, the paper and natural fibre layers are stacked in a mould in a previously determined sequence and then pressed in a hot press under temperature and pressure »**fig. 16**. A pulverous adhesion agent is evenly distributed between the layers. Mixing the cellulose/wood pulp fibres with water produces a pulpy mass. The pulp is deposited on the so-called wire, producing a paper web (paper matrix) with very high water content.

Mechanical properties of natural fibre-reinforced paper

Measurement results for unidirectional single layers (all fibres oriented in one direction) show that the tensile strengths in fibre direction are reached at approx. 180 MPa and a tensile modulus of elasticity of approx. 15 GPa. The tensile strength in the fibre direction is thus significantly higher than that of conventional papers, and even approaches that of aluminium alloys. The modulus of elasticity also reaches higher values than that of conventional papers but cannot compete with aluminium alloys (approx. 70 GPa) and steel (approx. 210 GPa).

When designing and dimensioning components made of anisotropic materials, it is critical to know the loads that will occur in the component so that the fibres are oriented in the direction of these loads. In the case of multi-axial loading, the reinforcing fibres have to be oriented in many different directions, which means that the mechanical properties of the resulting multi-layer composite are usually worse than if all fibres were aligned in one loading direction (i.e. a unidirectional web).

Free-formed components

The greatest demand for three-dimensionally shaped paper components has been in the packaging industry. Often, their geometries are limited to simple shapes with flat surfaces, produced by means of folding operations or drawing processes with a small drawing depth. Typical products include packaging boxes or paper plates. There has been increased research into three-dimensionally shaped paper components in recent years to expand the market potential. New processing methods and enhanced shaping options aim at opening the door for paper as a material for new purposes, for example, in the construction industry. Free-formed paper components for façades, as design elements or for trade fair construction are conceivable.

Various processes to produce three-dimensional paper products exist, many of which have been adopted from metal processing and adapted to the material and the resulting process requirements. A basic distinction is made between primary forming (moulding) and (de)forming processes.

In primary forming processes, the final geometry is created from a previously shapeless material. In terms of paper forming, this mainly involves pulp moulding – typical products are egg cartons or fruit trays. Pulp moulding means pouring or spraying a viscous fibre pulp into a mould and then drying it. Since the finished product is usually not pressed into the mould, the cast products have low strength and a rough surface. Another disadvantage is the high machine and tool costs that arise from the individual shapes of the formwork. Each product with a different shape requires a new formwork or a new tool, if not a new machine. The unbeatable advantage of pulp moulding, on the other hand, is the free geometries that can be produced as long as they comply with the design rules for moulding.

17 Conventional deep drawing of paper with a rigid tool.

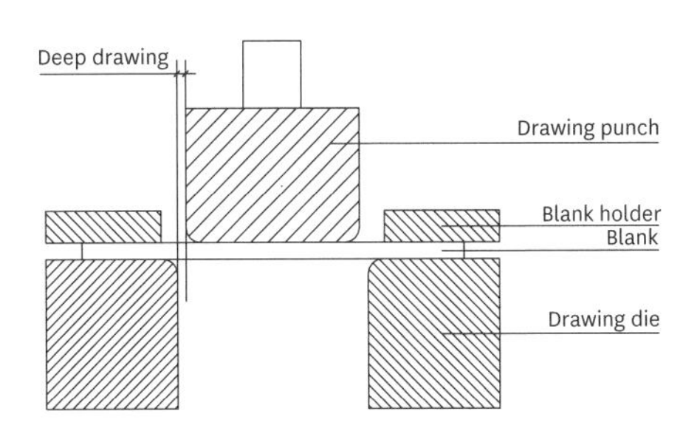

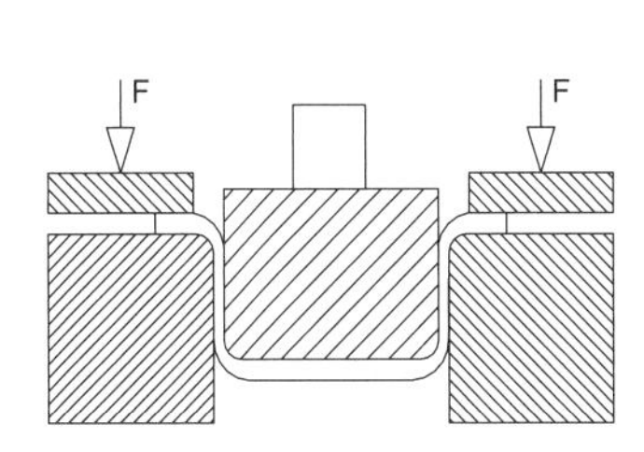

18 Squeeze moulding of paper. The material is pressed into a mould (drawing die) and thus brought into its final shape.

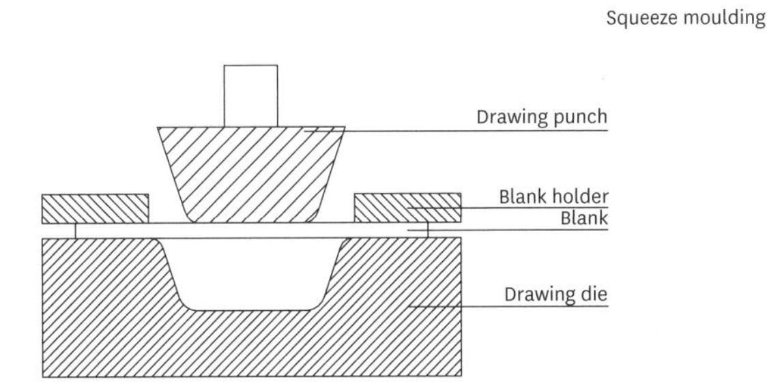

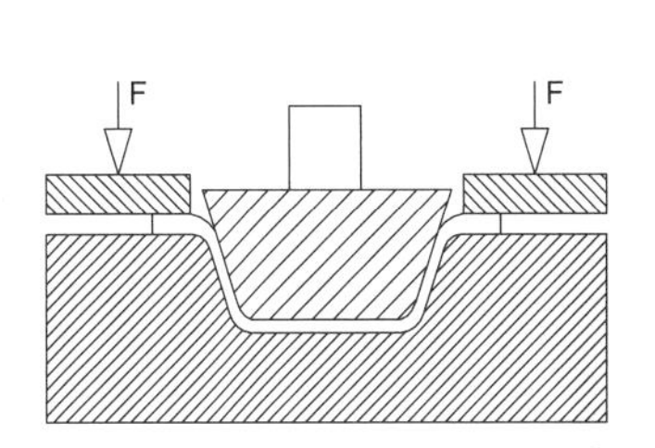

19 Active media-based deep drawing of paper.

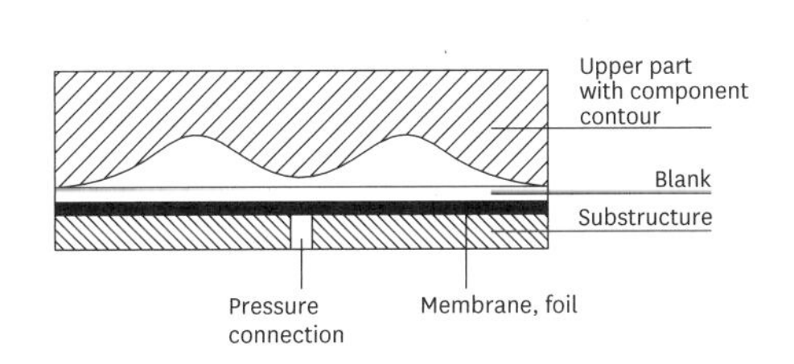

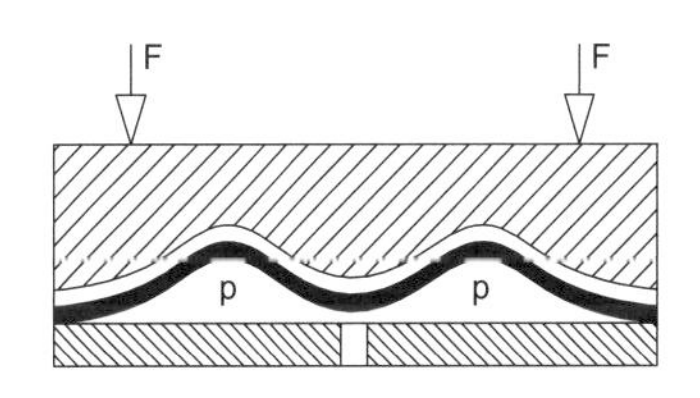

This contrasts with the (de)forming processes, which, above all, are characterised by low production costs, low cycle times and large quantities. In the forming processes relevant to paper forming, a flat semi-finished product is brought into its final, three-dimensional shape. Established processes to be used for forming paper are conventional deep drawing[7] active media-based deep drawing,[8] and squeeze moulding.[9]

In conventional deep drawing with rigid tools, a paper blank is formed using a punch with a predetermined punch path »**fig. 17**. The geometry of the finished cup results from the initial diameter of the blank, the diameter of the punch and the specified drawing ratio.

Unlike deep drawing, in squeeze moulding »**fig. 18**, the material is pressed into a mould ("die") and thus given its final shape. The final pressing of the material in the mould increases the shape retention of the geometry produced. However, the tooling costs are higher than with conventional deep drawing due to the required die

In active media-based deep drawing » **figs. 19, 21**, an active medium like oil or compressed air replaces the rigid tool punch. The active medium, usually compressed air, presses the workpiece to be formed into a die at high pressure. Since paper is porous and thus permeable to air, the workpiece must be separated from the active medium with a film or membrane.

Another forming process is incremental forming. For paper, this technique is still being researched. The following offers a first outlook on the method and its potential. With incremental forming » **fig. 20**, the material is only formed locally in a small area at a time. A tool follows previously programmed paths on the workpiece. After each path, the tool is adjusted radially and in the height direction so that any shape can be formed out of the material layer by layer without the need for expensive dies. Due to the long cycle times, incremental forming is not suitable for large batches. Instead, the process is aimed at customising and individually shaping workpieces.

However, when considering this technology, it should be noted that the stretching ability of paper is fundamentally different from that of other non-fibre-based materials. Under tensile load, paper stretches until the maximum elongation is reached and the paper tears. Under compressive and shear loads, paper wrinkles and individual layers of paper can delaminate. Unlike metallic materials, paper does not flow. However, research has shown that paper can be stretched further under multi-axial loading than is the case in material characterisation in the uniaxial tensile test. Incremental forming thus promises potential for free-formed components. Incremental shaping of paper is particularly well suited for unusual surface designs.[10]

20 Incremental forming of paper parts.

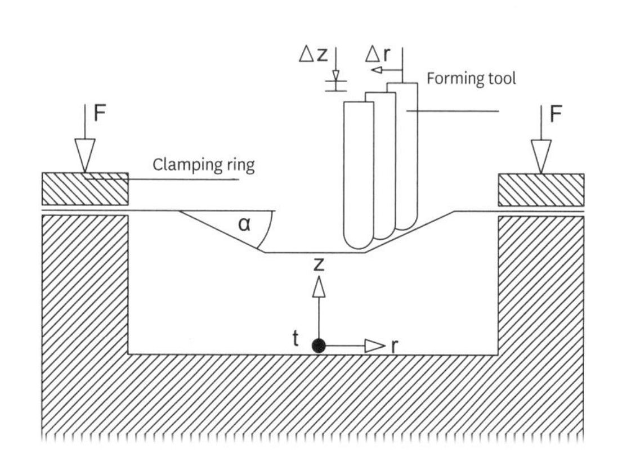

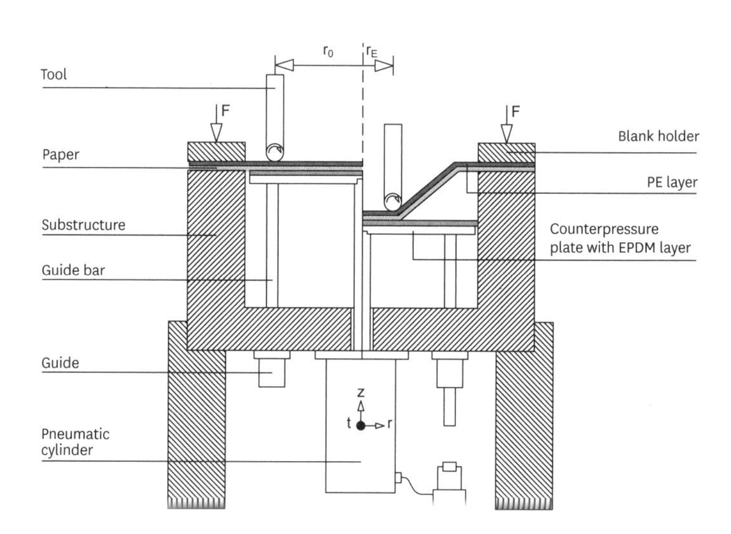

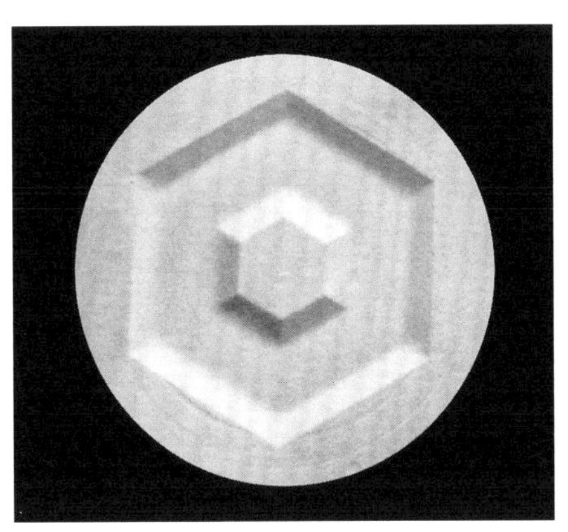

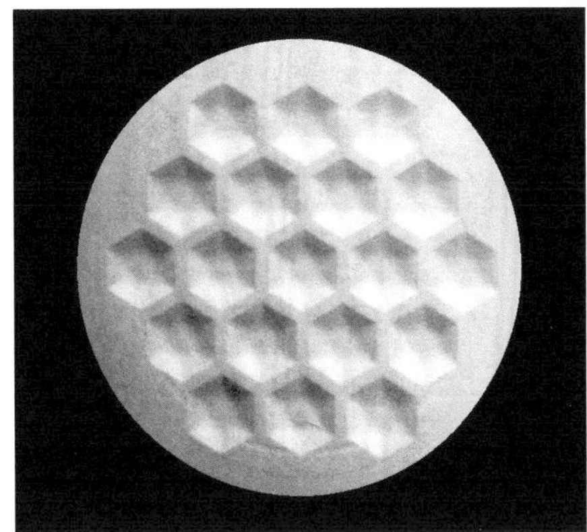

21 Results of active media-based forming.

REFERENCES

1 *DIN 6730, 2011-02: Paper and Board – Vocabulary*, Berlin: Beuth, 2011.
2 Erwin Bachmaier, Bundesverband Druck und Medien (ed.), Ausbildungsleitfaden Druckweiterverarbeitung. Werk- und Hilfsstoffe, Berlin, 2017, https://www.mediencommunity.de/content/31-papier-karton-pappe, accessed 20 March 2019.
3 *DIN 6735, 2010-2: Paper, Board and Pulps – Survey of Terms and Definitions*, Berlin: Beuth, 2010.
4 Verband der Wellpappen-Industrie [illegible] *pappe sicher verpacken und erfolgreich verkaufen*, 2007.
5 Stephan Schütz, *Von der Faser zum Haus. Das Potential von gefalteten Wabenplatten aus Papierwerkstoffen in ihrer architektonischen Anwendung*, 2015, Dissertation, Bauhaus-Universität Weimar, https://e-pub.uni-weimar.de/opus4/frontdoor/index/index/docId/3804, accessed 13 June 2019.
6 *DIN 55468-1, 2015-6: Packaging Materials – Corrugated Board – Part 1: Requirements, Testing*, Berlin: Beuth, 2015.
7 Marek Hauptmann, *Die gezielte Prozessführung und Möglichkeiten zur Prozessüberwachung beim mehrdimensionalen Umformen von Karton durch Ziehen*, Dissertation, Technische Universität Dresden, Institut für Verarbeitungsmaschinen und Mobile Arbeitsmaschinen, 2010.
8 Dominik Huttel, Peter Groche, "New Hydroforming Concepts for Sustainable Fiber Material (paperboard)", in: International Conference New Developments in Hydroforming, 2014, pp. 1–10.
9 Ville Leminen, Panu Tanninen, Petri [illegible] Effect of Paperboard Thickness and Mould Clearance in the Press Forming Process", in: *BioResources* 8(4), 2013.
10 Dominik Huttel, *Wirkmedienbasiertes Umformen von Papier*, Dissertation, TU Darmstadt, Institute of Production Engineering and Forming Machines, 2015.

4 BUILDING CONSTRUCTION

This chapter deals with the structural fundamentals of paper structures. First, the semi-finished products and components previously described in chapter 3 are combined into idealised structures, and their mechanical properties are discussed. This is followed by a presentation of joining techniques that can be used to connect these structures. The chapter concludes with construction typologies derived from the previous parts. More in-depth knowledge of the mechanics of paper and related calculation methods can be found in the chapter “Facts and Figures for Engineers” at the end of the book.

Idealised structures

Paper is produced in the form of a paper web and thus as a flat, orthotropic material. The semi-finished products and components produced in subsequent production steps are suitable for constructing user-specific load-bearing systems and building structures with relevant mechanical properties. For efficient material use, these properties must be taken into account when selecting materials for structural design **»chapter 5, pp. 68–73.**

For an analysis of the construction typologies and the resulting structural systems, these components can be summarised into the following basic idealised structures:

- Linear elements
- Surface elements
- Volumetric elements

These elements are used to discretise the structure to perform both analytical and numerical calculations. The following section describes the basic mechanical properties of the aforementioned idealised structures **»fig. 1.**

Linear elements

Beams

In linear elements, the length of the element dominates over the width and the height of the cross-section. In structural analysis, linear elements are used to calculate the acting forces and moments, the so-called internal forces **»fig. 2.**

Classic linear elements include cardboard tubes and cardboard profiles **»sections “Tubes and cores” and “Profiles”, pp. 42–45 and 45–46**

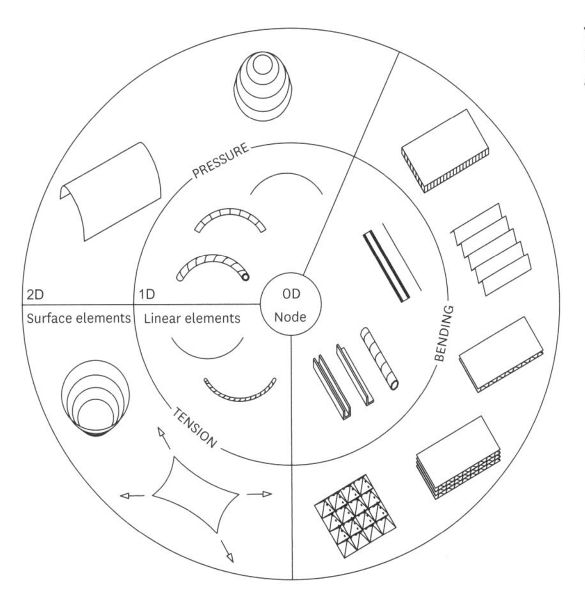

1 Design principles with idealised structures according to load type.

2 Beams with internal forces (2D and 3D) where S means shear force, M moment and N normal force.

Spiral wound tubes will twist if subjected to pure compressive load due to the spiral winding method of the roll core. Torsion occurs along the longitudinal axis of the element, creating a normal force-torsion relationship when considering the tube as a linear element. If, for example, a tube serves as a load-bearing component and is loaded with pressure, the support head gets twisted. As a result, bending moments are introduced into adjacent horizontal beams, which must be taken into account when dimensioning the overall load-bearing system.

Next to cardboard tubes and cardboard profiles, linear elements such as beams can be assembled from multiple components. A solid beam, for example, can be formed by laminating solid cardboard sheets. It should be noted that with such beams, shear deformation occurs in addition to bending deformation. Tests have shown that shear stresses can lead to failure in the adhesive layer. The adhesive layer gives way, and the solid cardboard sheets detach from each other and shift. Such beams are therefore not suitable for structural applications.

Another approach is to resort to relatively thin-walled paper components and use, for example, the intralaminar, i.e. material's own shear, stiffness. An example of this method is a thin-walled closed box girder.

Cables

The cable is a special case of the beam; its global bending stiffness is very low, and normal tensile forces are transmitted as a matter of principle. Loads can also act vertically on a cable, but only when it is tensioned.

Industrially produced paper cables with a diameter of 11.5mm, for example, have a tensile strength of over 2000N (approx. 200kg as weight/vertical force). Experiments have shown that these cables allow for very large elongations during initial loading (> 3%), making it difficult to prove serviceability. Pre-stretching might remedy this problem.

3 Schematic representation of a plate as a surface element with the relevant directions of force and moment.

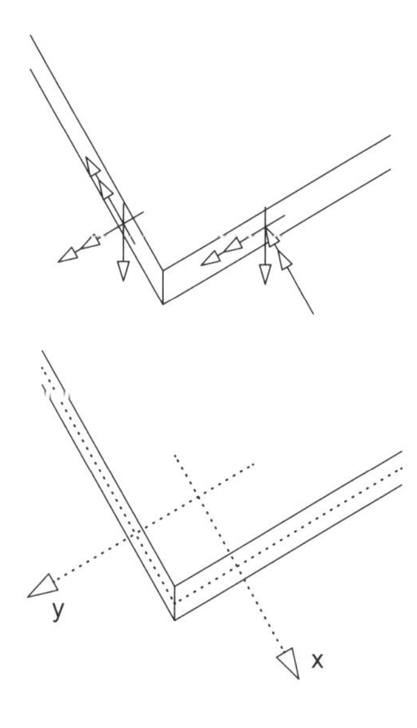

Area elements

Area elements include plates, diaphragms, shells and membranes. These are described in the following sections.

Plates

Plates are flat elements that are predominately loaded perpendicular to the surface plane »**fig. 3**. Accordingly, they transmit forces perpendicular to the surface plane and moments around the axes in this plane. The central area of the plate serves as an idealised structure for the consideration of mechanical values such as forces, moments and stresses.

In terms of mechanic properties, their structure makes paper-based honeycomb panels »**section "Sandwich structures", pp. 40–42**, count as panels. They transfer corresponding loads very efficiently. Accordingly, the static height of the honeycomb cores can accommodate large bending moments despite relatively thin cover layers. However, resulting shear loads should not be neglected. The shear resistance, i.e. the ability to withstand the load, is determined not only by the paper material but also by the cell wall geometry and the adhesives selected. Due to the thin cover layers, these materials (with honeycomb cores) are mainly suitable as plates rather than diaphragms.

Diaphragms

Unlike plates, diaphragms are loaded in the plane, i.e. parallel to the defined surface »**fig. 4**. In-plane loads create stresses that remain approximately constant across the thickness. Both normal stresses σ_{xx}, σ_{yy} and shear stress τ_{xy} occur in the x-y plane.

In the case of a beam whose height and length are of the same order of magnitude, we need to assess whether the beam theory can still be applied or whether the system at hand is more subject to the plane stress theory.

In multi-layer structures, the stiffness of the individual layers determines how the load is distributed in those individual layers »**fig. 5**. As a general rule, it can be assumed that, with identical deformation, elements with higher stiffness can also absorb higher loads.

In the case of honeycomb or multi-wall corrugated boards, this behaviour can be well illustrated with an edge crush test (ECT). It shows that mainly the surface layers are stressed, and the flexible core escapes the load »**fig. 6**. Since the face layers of honeycomb or multi-wall corrugated boards are typically very thin, these boards are unsuitable for transferring in-plane loads in a load-bearing system.

Corrugated board is quite different: corrugated boards acting as diaphragms transfer high loads, especially parallel to the flute direction. Since the walls of corrugated cardboard boxes are oriented accordingly, i.e. parallel to the flute direction, they allow high filling weights and can be stacked easily. The effective depth, which provides the stability of the corrugated board, results from the corrugated core. Even stronger are triple-wall corrugated boards, commonly used due to further improved properties.

4 Schematic representation of a diaphragm as a surface element with the relevant load directions.

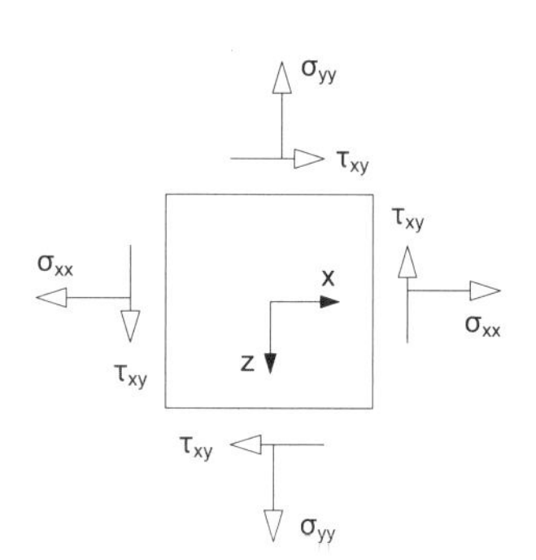

Shells

Shells are three-dimensionally curved-area elements. Shaped intelligently, these often very thin-walled constructions can carry loads very efficiently. The category of shell

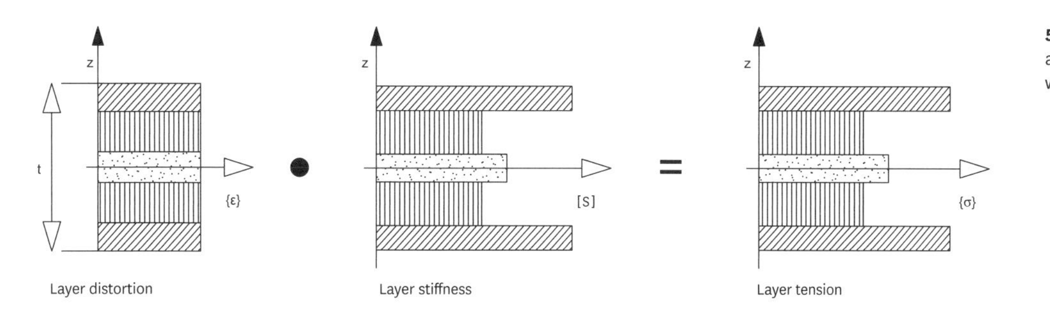

5 Strains and stresses of a multi-layer structure with different stiffness.

elements includes the three-dimensionally shaped elements described in the **» section "Free-formed components", pp. 51–53**, which cannot be classified as linear or other area elements.

6 Compression tests on honeycomb boards and loading directions (FCT: flat crush test, ECT: edge crush test).

ECT 0
ECT 30
FCT

Membranes

Membranes are flat, flexible load-bearing elements that absorb in-plane tensile forces. They thus form a 2D counterpart to the cable element. Membrane structures can transfer forces orthogonally to the plane of the surface if tensile stresses prevail in the plane.

Volume elements

Volume elements are solid, three-dimensional, block-like structures. In principle, no information can be given about the resulting internal forces at their cutting edges. Instead, the stresses that arise are assessed using numerical methods.

Joining techniques

The most important joining techniques for the previously described idealised structures are explained below to show how broad the spectrum of possibilities is for developing assembly systems and construction details for paper constructions.

Currently, there are no standards for connections in and of paper structures. Thus, joining techniques from other fields most suitable for paper constructions serve as a basis **» fig. 7**. These are grouped into the four main groups adhesive joints, flexible joints, mechanical fixations and positive (form-fit) joints. Concepts for implementing these methods in different paper constructions are described below.

Adhesive joints

Bonding or adhesive joints can be accomplished by laminating, taping or couching. Lamination is probably the most common method for joining paper-based materials, both in manufacturing components and post-processing to form subsequent assemblies. The adhesives differ depending on the type of laminate: PU (polyurethane) and PVA adhesives are preferred for structural purposes because they produce high-strength joints. Starch glue, on the other hand, is more environmentally friendly but does not achieve such strength.

7 Basic principles of different joining techniques.

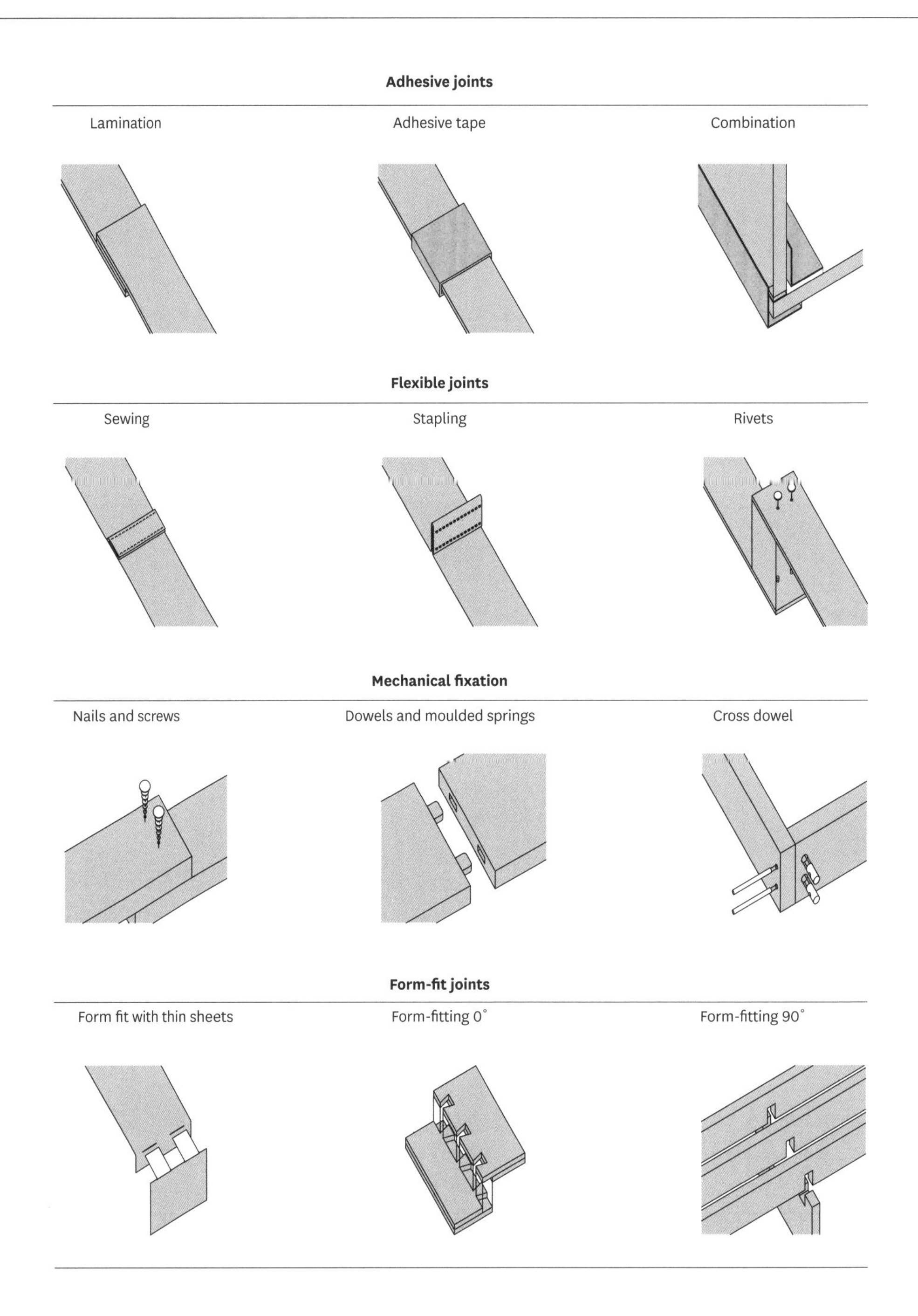

Similarly common is bonding with adhesive tape, preferred for rapid prototyping and reversible assemblies. Single-sided adhesive tape is recommended for sealing a joint between two elements, especially for temporary installations. Double-sided adhesive tape can be used to create tight lap joints.

Couch pressing is a traditional method for joining individual layers of paper under pressure and high humidity. It is primarily used as part of the production process. In principle, the paper layers are moistened and pressed together until the fibres bond in the contact surface area between the sheets, after which the material must dry. This joining technique might be used for building with paper, but it is not yet fully developed for the purpose.

Flexible joints

Flexible joining techniques are mainly suited for thin sheets and panels that can be both rigid and flexible, similar to the joining technique of fabric structures. These methods are particularly popular for thin-walled, tension-loaded constructions. Joining is done by sewing or stapling and by means of snap connectors, such as mini-rivets and grommets.

It is not only thin sheets of paper that can be sewn together in a similar way to textiles »**fig. 7**; thin-walled tensile structures can also be joined in this way, as the following consideration shows: enveloping a construction made up of individual panels with a top layer creates a homogeneous surface finish. At the same time, this skin can hold the construction together by pressure if the seams are sealed with glue. However, such seams can fail, for example, due to delamination. To prevent such failure and ensure a durable seam, the paper sheets can be additionally bound with stitches. The combination of tacking and glueing is another possibility.

A variety of different stitching methods are available for sheet materials of different qualities. The thickness and the matrix of the material layer determine the density of the seam. To avoid breakage along the seam, the stitch length should be shorter for thicker layers and longer for thinner layers. Materials joined with this technique should not be thicker than 3mm, depending on the hardness of the material. When tacking layers of paper, the most resistant direction of the paper should be used to absorb tensile forces, combined with seams oriented at right angles to it.

Snap connectors, which can be attached at specific points, ensure good anchoring in the material. They come in different sizes depending on the thickness of the material they need to bridge. Grommets form the basis for fixing and essentially serve to create tear-proof holes through which a cable can be pulled and tensioned. Mini-rivets are suitable for fastening surfaces together with a lap joint. To reinforce the material in specific areas, highly functionalised fibres are integrated into the matrix, either during the production process or at a later stage in the form of patches.[1] If access to the joint is limited, blind rivets can be used; thin sheet metal is particularly effective for joining. Blind rivets can be easily replaced, but often at the expense of the material as the rivets are usually completely damaged when removed.

Mechanical fixation

The joining techniques described below are widely used in timber construction and can largely be transferred to paper constructions, provided that the applicable standards are adapted accordingly. The various fastening methods range from minimalist solutions to sophisticated fastenings. Selecting the appropriate joint type depends on the properties and characteristics of the building component, i.e. its density, structure (hollow or solid), surface finish, etc.

Solid panels with depths of up to 14mm can be joined with a seam made of staples. Nails should be used in the case of thicker joints subjected to shear forces. If the joints are subjected to tensile forces, steep thread screws should be used instead. Ordinary metal screws are only suitable for solid components with high density. Coatings or wood blocks can serve as local reinforcement for the connection of softer paper components with metal screws. The use of plastic screws with an extremely steep thread is limited to compression-loaded lightweight panels with cavities, as their efficiency under tensile stress is quite limited.

Another strategy is to apply pressure plates to the surfaces to protect the openings and distribute the loads optimally across the surface. An ideal solution for connecting beams with bolted joints is shown in the example of the Japanese Pavilion » **chapter 6, pp. 132–133.**[2] When panels are bolted together, this is often combined with form-fit techniques to eliminate shear forces. Similar methods are used with solid rivets.

Dowels and moulded springs can be used to align components with each other precisely, but they do not transmit any forces as this is only effectively possible with more solid materials. Therefore, many projects employ a strategy of using wood blocks as local reinforcement because the blocks allow for force transmission. An easier way to fix the components in place is to use modern cross dowels, which are familiar from furniture construction.

In terms of the structural performance of the various fastening elements, initial findings for a number of tested designs already exist but further systematic experiments are required to obtain values that can be regarded as standard. Therefore, for the time being, the standards of Eurocode EN1995-1-1 on the design of timber structures – especially section 8, Connections[3] – are considered a useful guide. As far as dimensioning is concerned, the guidelines contained in the standard provide some orientation, as, with paper-based components, higher minimum distances between the joining elements are generally to be expected.

Form-fit joints

Form-fit construction methods » **fig. 8** are familiar from packaging products, such as boxes made from folded cardboard. They are particularly useful as complementary joining methods when building with paper, in combination with either glue or mechanical fasteners. Relying solely on friction or force-fit positive locking connections can be problematic with paper-based materials because they tend to lose stiffness over time. Form-fit joints are an easy way to create lap joints between members or to assemble "puzzle-shaped" façade panels. The construction in » **fig. 8 c** is an exception ("dry form-fit"). In critical areas, designers should consider curved edges as much as possible to optimise stress distribution. The plug-in hybrid connections discussed below also belong to the category of form-fit connections.

8 Design examples of connections a. braided lap joint, b. slanted form-fit joint, c. waffle construction

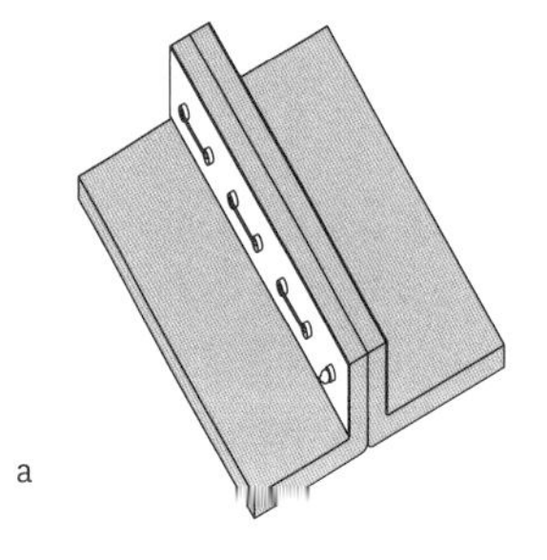

a

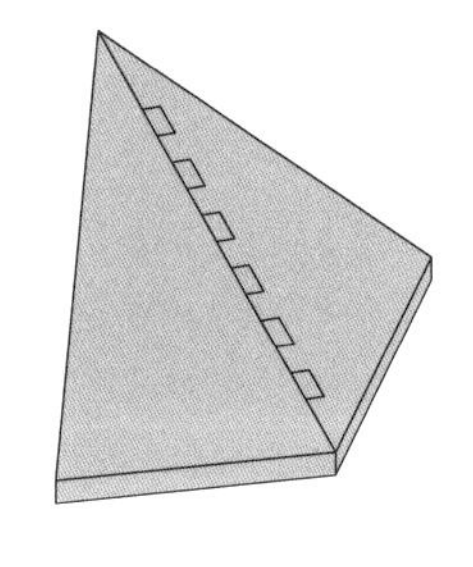

b

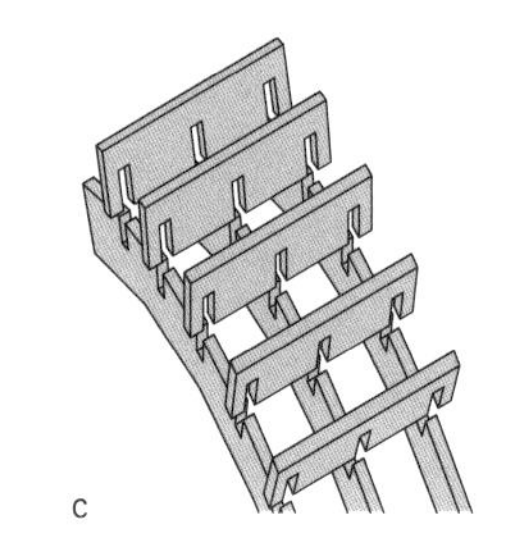

c

Connections for frame constructions

In the context of the connections for structural frames explained in this section, the load-bearing structure is considered as being made up of cardboard tubes. However, such approaches can also be found in common tubular, steel and bamboo structures.

Among the simplest structural systems are linear assemblies, divided into the three variants of internal sleeves, external sleeves and clamp joints **»fig. 9**. A non-visible inner tenon reinforces the system in case of bending loads, especially if it is made of solid material. An outer tenon, in turn, covers the joint between the members and thus protects the most important construction elements. With paper-based materials, press-fit concepts have the previously mentioned disadvantages of pure positive locking connections, and they require zero tolerance, which often poses problems during assembly. In contrast, clamp joints are much easier to mount and ensure a firm connection.

Multi-axial connections based on the same connection methodology can also be divided into three main categories: form-fit joints, press-fit joints and sleeve joints **»figs. 10–12**. In contrast to pure form-fit connections, press-fit and sleeve joints require suitable interconnectors made of common construction materials.

In the examples of form-fit connections shown in **»fig. 10**, fittings or lugs exert pressure on the members to distribute the loads across the contact surfaces. One significant advantage of these solutions is the simplicity of production and assembly. However, the strength of such connections is limited because the recesses in the members weaken them considerably. Therefore, they are preferred for smaller, temporary structures, such as the Paper Log House **»chapter 6, pp. 88–89**.[4]

Press-fit connections are usually made of wood-based materials. A distinction is made between two principal versions: on the one hand, there are light press-fit connections formed with slender, interlocking wood elements **»fig. 11**. They are particularly prevalent in traditional Japanese timber construction and were also used in House 01 **»chapter 6, pp. 120–121**.[5] As these connections are relatively easy to manufacture, they are often preferred over the various variants of solid press-fit connections. However, the contact area between the node and the tube is usually limited, which can lead to structural problems.

Solid press-fit joints can be formed with wood blocks joined together to make a knot. This method allows fasteners to be better fitted to the cardboard tubes, maximising the contact area between the structural elements. The more precisely the components are manufactured, the better the load-bearing capacity.

In both variants, the structural integrity strongly depends on the chosen wood species. In addition, minimal tolerances can improve load-bearing capacity and integrity, but this involves a very labour-intensive and time-consuming assembly process.

Steel connectors, combined with pre-tensioning of the cardboard tubes, fulfil higher structural requirements. In this case, the metal tenons of all adjacent members concentrate in a central articulated joint.

In the case of sleeve joints, moulded parts made of aluminium or steel usually encase the cardboard pipes and are fastened with compression rings or clamps; rivets; bolts; or, more recently, composite materials **»fig. 12**.

All the connections explained here have different qualities. However, the differences between high-performance metal connections and the other connection types are small. Softer materials are advantageous for handmade solutions, and it makes just as much sense for the designer to be closely involved in the manufacturing process. Joints made of wood-based materials or composites manufactured from readily

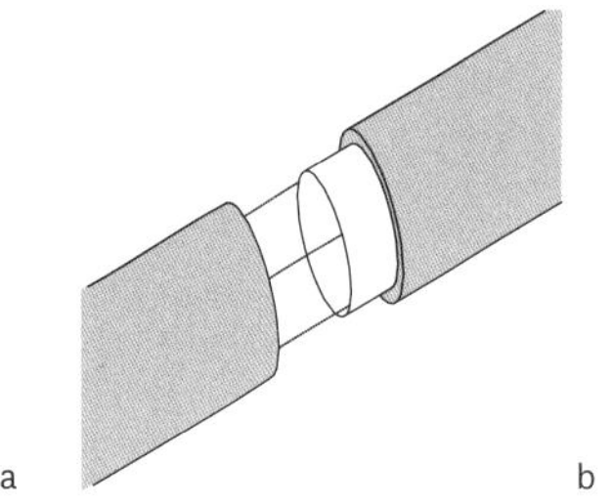

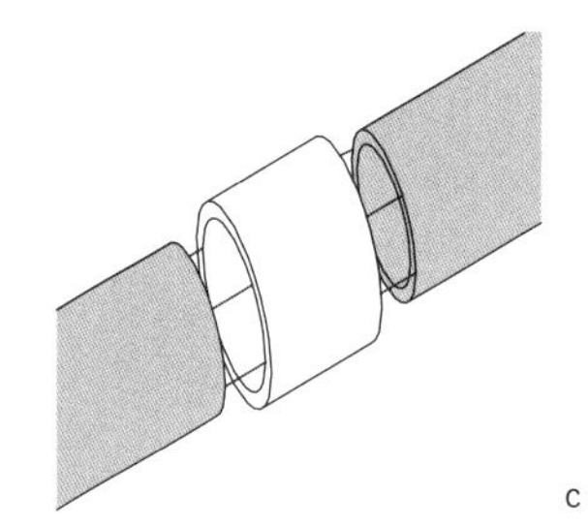

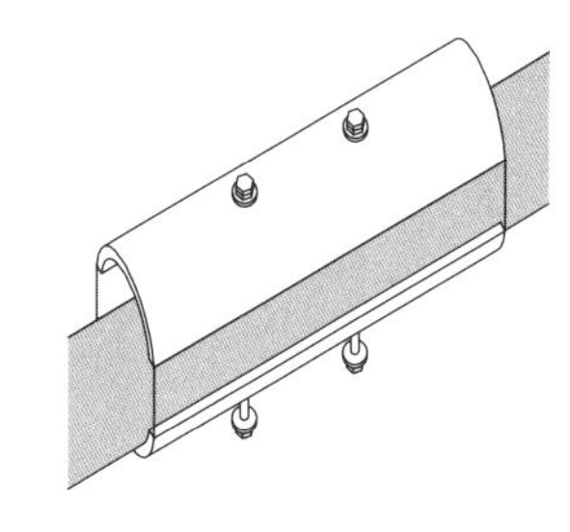

9 Linear joints for cardboard tubes: a. inner sleeve, b. outer sleeve, c. pressure profile/clamp.

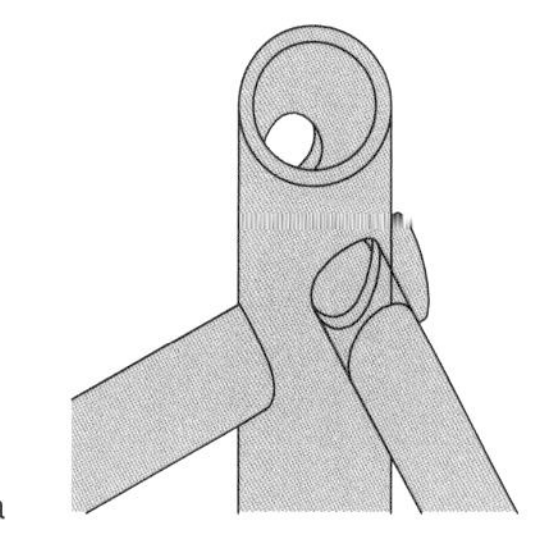

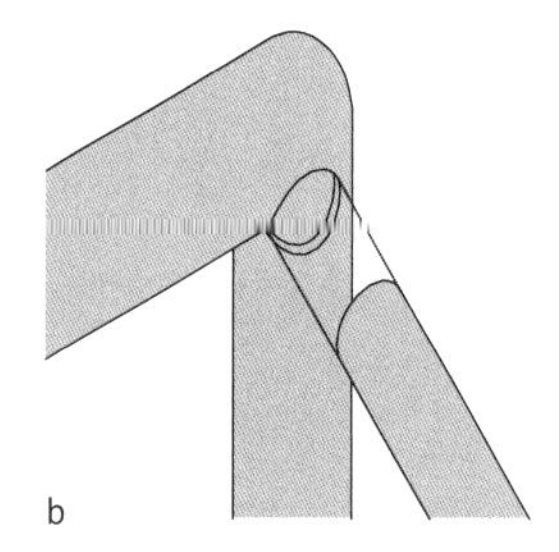

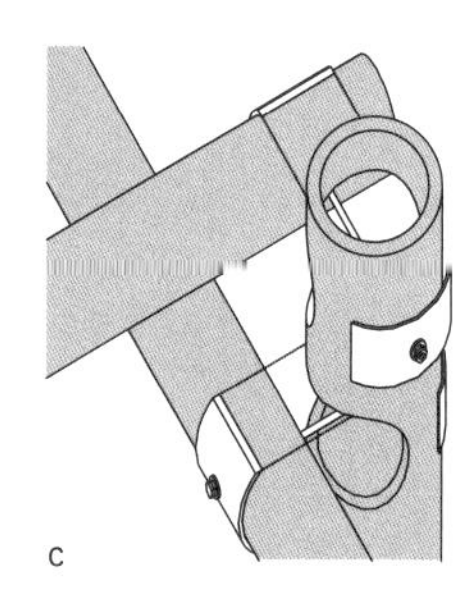

10 Form-fit connections for cardboard tubes: a. complete form-fit connection, b. one-sided form-fit connection, c. frictional connection.

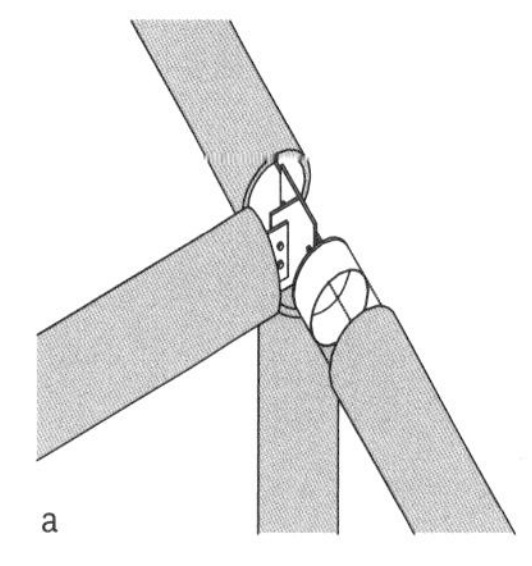

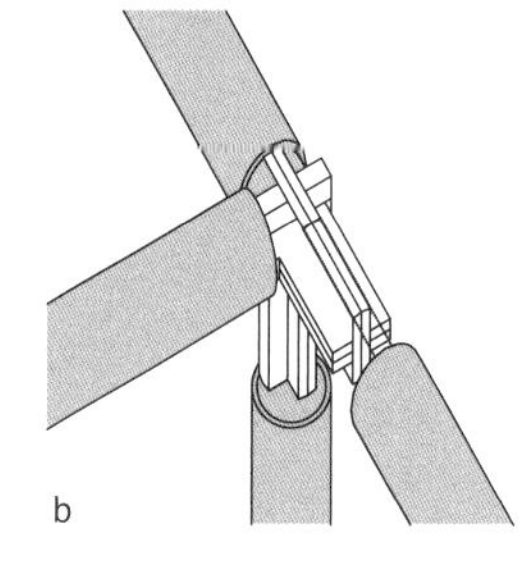

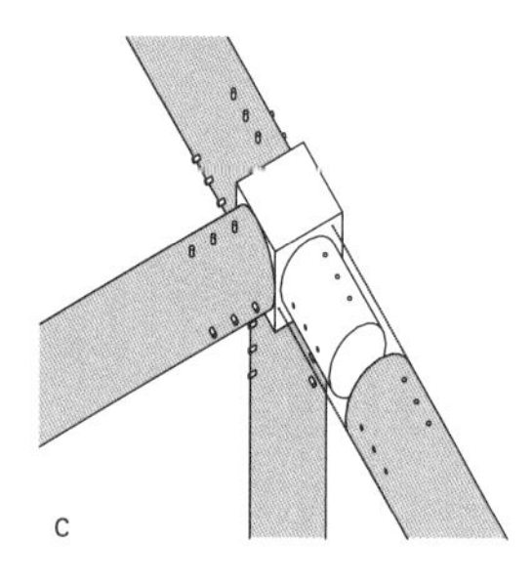

11 Press-fit connections for cardboard tubes: a. metal tenon, b. light cross tenon, c. tenon with full contact.

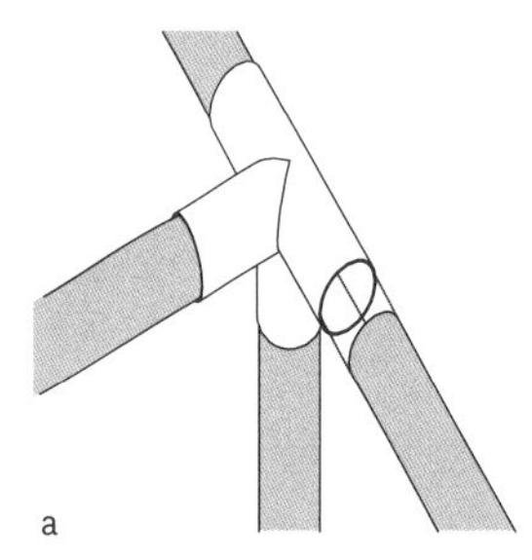

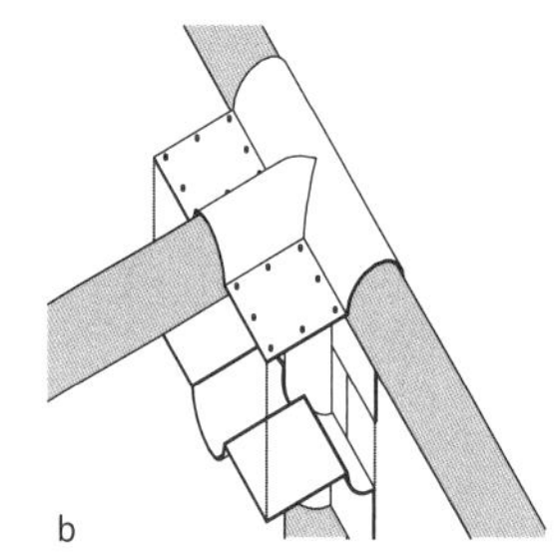

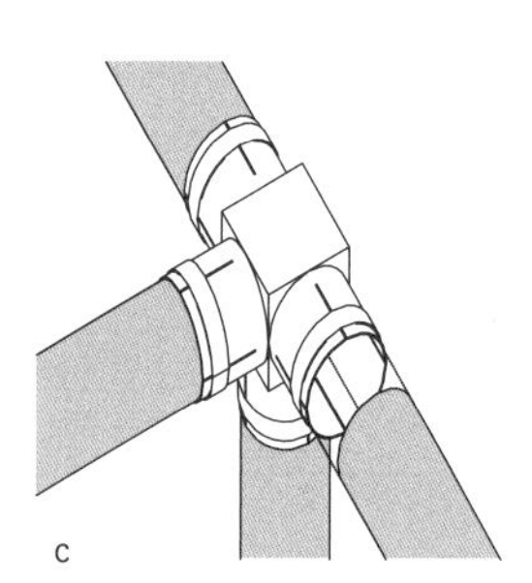

12 Sleeve joints for cardboard tubes: a. simple positive connection, b. with pressure plates, c. with interconnectors.

available fibre products are particularly suited for paper-based constructions. Relevant research projects focusing on detailing, prototyping, and load-bearing tests provide insightful information.[6]

Construction typologies

The range of construction typologies for building with paper is very broad but can generally be divided into the three categories of plate, shell and frame construction »**fig. 13**.

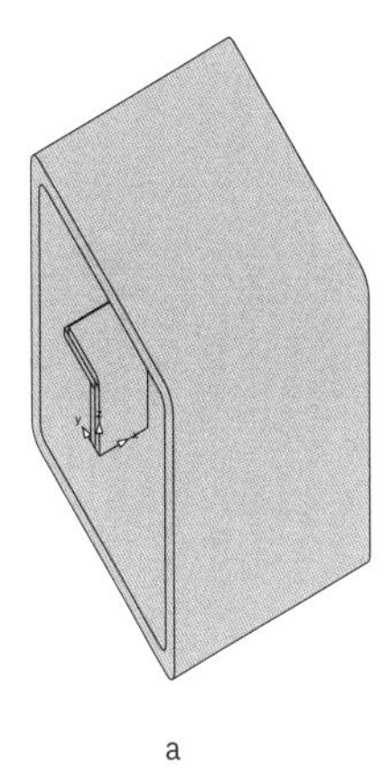

a

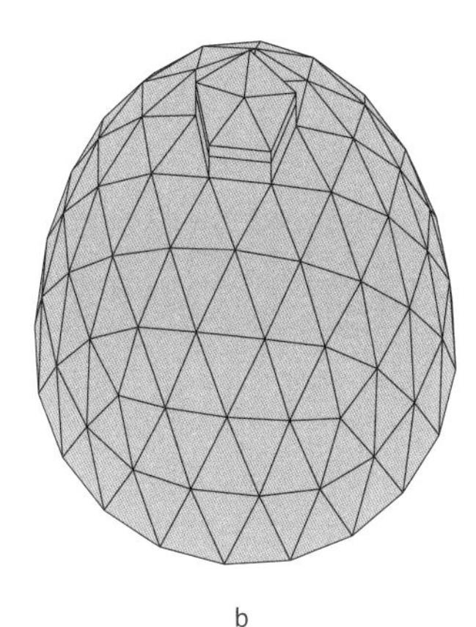

b

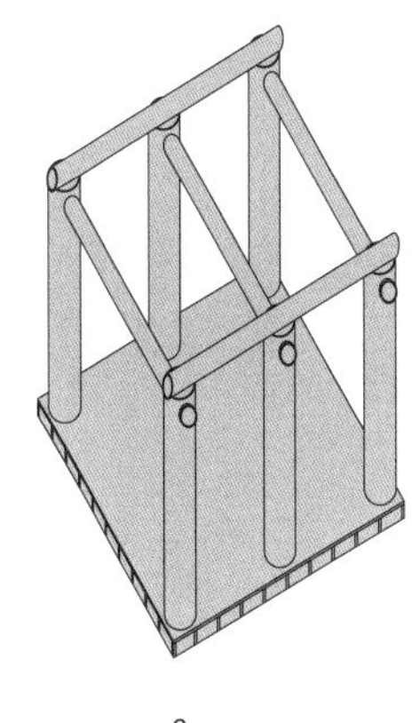

c

13 Main construction typologies for building with paper: a. plate, b. shell, c. frame.

In principle, the material lends itself to free-standing, ground-level structures designed for temporary use.

Plate and shell constructions are mainly composed of multi-layer composite load-bearing panels – depending on the specific function of a particular wall (part of an exterior or interior wall, the roof, etc.) and the general requirements for the construction. Representative examples of functional wall elements are listed in **»chapter 5, p. 81**.

Plate construction

The vertical (z) and perpendicular (y) stiffness of the panels to the main plane is an essential factor **»fig. 13a**. Their structural stability can be increased by strategically integrating different paper products into a multi-layer sandwich panel. Solid boards are particularly suitable and popular for this purpose. The advantages of honeycomb panels are noticeable in the case of forces acting at right angles on the surface (e.g. wind loads).

The decisive factor is the composition of the individual layers. This will be illustrated by other essential aspects that show how the design of a panel can be changed to suit a particular purpose. The intermediate layers, which run parallel to the panel's main plane, bear most of the axial loads and contribute to the bending resistance. This aspect, as well as the general stability requirements, determines the number and thickness of such solid layers and the integration of lighter layers, which can be employed to act as a buffer zone for climate regulation. Therefore, the total thickness of a panel can vary considerably, depending on the priority of the previously mentioned decisions and the building-physical properties of the chosen materials. Other aspects that need to be considered during planning and can contribute to both stability and durability are the quality of the outer layer or surface finish, the edge finish and the tolerances between the individual panels. Also important is the hardening of the outer surfaces, rounded outer edges and seals to compensate for tolerances and repel external influences such as rainwater.

Plate and shell constructions are predestined for prefabrication, especially because the lamination process requires certain conditions and special machines to [illegible]

14 Thin-walled structures,
a. origami structure,
b. tension-based structure
with narrow frame
construction.

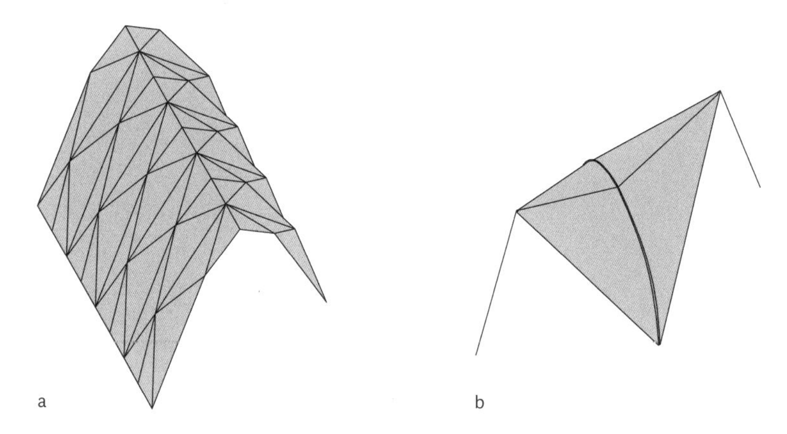

considered the most recommendable design and assembly concepts, not least because they can be easily relocated. One good example of a modular prefabricated structure is the Wikkelhouse »**chapter 6, pp. 90–95**).

Shell construction

Certain shapes with closed geometries can increase load-bearing capacity and structural integrity by applying pressure to the structural elements and guiding wind loads around the building with aerodynamically shaped façades. In this respect, the most common forms include arches and dome structures, which are the preferred choice for temporary installations and demonstration projects and often feature innovative designs, such as cellular cavity structures with hollow cardboard panels.[7] Fabrication is a critical aspect when the construction is made up of many differently shaped panels. Therefore, the design geometry should be optimised accordingly, and the number of geometrically different panels should be kept to a minimum.

Origami constructions are a special type of shell construction »**fig. 14a**. Many temporary emergency shelters are based on this construction method inspired by the traditional Japanese art of folding.[8] Paper products are ideal for thin-walled folded constructions because the material is inherently easy to fold.[9] Folding patterns can create special stability effects. However, since folded edges are susceptible to damage, the durability of origami constructions is limited. But if these edges are reinforced, the service life can significantly increase. In this respect, structures based on tensile forces with narrow frame constructions are also interesting »**fig. 14b**. Such tent-like structures, e.g. for shelters, exploit the advantage that paper products can absorb tensile forces.

Frame construction

Frame constructions realised with cardboard beams offer another solution for load-bearing structures clad with lightweight materials. Constructions based on an orthogonal grid seem particularly advantageous because the spaces they create can be optimally utilised – for example, with regard to furnishing. However, even with this typology, arch and dome constructions offer structural advantages, especially for larger structures such as temporary theatres or other event buildings. This advantage must be weighed against the poorer space utilisation.

There are various options for the primary load-bearing components. These include profiled beams such as cardboard tubes or multi-layer composite beams made of cardboard. Depending on the material, there are specific areas of application: profiled beams are better at absorbing compressive forces, while composite beams are better at resisting bending forces.

Frame constructions made of cardboard tubes are extremely popular, as the many examples in **»chapter 6** show. The Paper Log Houses by Shigeru Ban are based on small, differently configured frame constructions. They demonstrate their suitability for temporary housing projects such as emergency shelters **»chapter 6, pp. 88–89**. The cardboard tubes are often pre-tensioned to increase stability – a strategy that was also chosen for constructing the Paper Theatre in IJburg **»chapter 6, pp. 134–135**, and is still considered an effective method of permanently counteracting the material's tendency to give way.[10] In addition, the relative flexibility of the material points to the production of multi-layer gridshell structures, as in the case of the Japanese Pavilion by Shigeru Ban **»chapter 6, pp. 132–133.**

Most examples of composite girder structures use multi-layered cardboard elements, such as the Apeldoorn Theatre **»chapter 6, pp. 130–131**, by Hans Ruijssenaars,[11] or honeycomb components, such as the Nemunoki Children's Museum by Shigeru Ban. The performance characteristics of the latter can vary depending on the composition of the core. In the Nemunoki Children's Museum, the load-bearing structure of the roof is made of corrugated boards arranged in multiple layers in a honeycomb pattern to achieve high load-bearing capacities and bending stiffness.

Generally speaking, the structural grid and the achievable span depend strongly on the type of support element. Research institutes conduct structural design tests to provide insightful information for some of these components. Information on the performance of cardboard tubes can be found in previous research,[12] in the case of Shigeru Ban in work monographs[13] and current research results.[14]

With regard to composite members, other relevant studies deal with beams made of honeycomb panel assemblies or a combination of cardboard tubes,[15] as well as the forming of beams from folded honeycomb boards.[16]

Hybrid constructions

Different construction principles are often combined in a hybrid construction to increase the load-bearing capacity and improve the functionality of the building. Beams or slats **»fig. 15**, for example, can be used to reinforce the building envelope of panel and shell structures and to create spaces to integrate essential building services elements. Conversely, with frame structures, the cladding can brace the supporting structure. Such examples of hybrid construction typologies can be seen in research projects like House of Cards **»chapter 6, pp. 114–117**[17] or the very first prototype of a complete house developed as part of the BAMP! project under the name House 01 **»chapter 6, pp. 120–121**. The combination with other building materials (for example, plastic films or coatings) is also interesting, exemplified by the cardboard building of the Westborough School **»chapter 6, pp. 96–99.**[18]

One important structural aspect is the foundation since it is here where the load-bearing structure is anchored to the ground. There are a few basic foundation

a

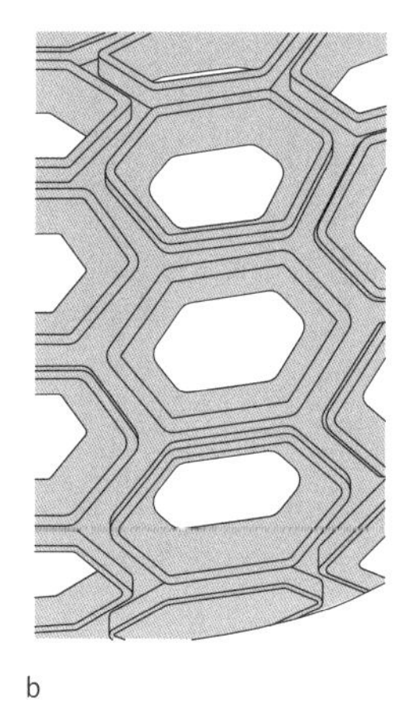
b

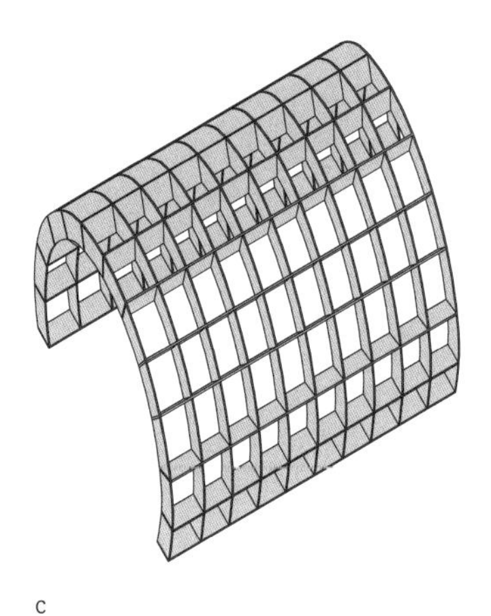
c

15 Lamellar paper structures: a. rhombic, b. honeycomb-shaped, c. rectangular.

strategies: for short-term installations, a base plate in the form of a heavy platform is usually structurally sufficient and easy to dismantle. Such a platform is often a hollow base filled with heavy natural materials such as sand or gravel. Concrete foundations are options for more permanent constructions – either in the form of a continuous slab underneath the entire construction or divided into several individual concrete bases. For smaller constructions, parallel beams resting on the ground are sufficient to raise the cardboard construction above the ground, as in the case of the Wikkelhouse **» chapter 6, pp. 90–95.**[19]

Frame constructions can be founded on steel foundations, or, more precisely, steel elements can be mounted as a connection between the structure and the concrete foundation. Wood fittings can be an alternative for constructions made of cardboard tubes. In-principle solutions for construction details are shown in **» chapter 5, section "Typical construction details", pp. 80–83.**

REFERENCES

1 Jan Knippers, Jan Cremers, Markus Gabler, *Atlas Kunststoffe + Membranen: Werkstoffe und Halbzeuge, Formfindung und Konstruktion*, Munich: Edition Detail, 2010.
2 Matilda McQuaid, *Shigeru Ban*, London: Phaidon, 2003.
3 *DIN EN 1995-1-1:2010-12, Eurocode 5: Design of timber structures – Part 1-1: General – Common Rules and Rules for Buildings*, Berlin: Beuth Verlag, 2010.
4 Heidi Zuckerman Jacobson, *Shigeru Ban – Humanitarian Architecture*, Aspen: Aspen Art Museum, 2014.
5 Evgenia Kanli, Rebecca Bach, Robert Götzinger, Nihat Kiziltoprak, Ulrich Knaack, Samuel Schabel, Jens Schneider, "Case Study: Development and Evaluation Methods for Bio-Based Construction Realized with Paper-Based Building Materials", in: 3rd International Conference on Bio-Based Materials, ICBBM, Belfast, RILEM Publications, 2019.
6 Evgenia Kanli, *Experimental Investigations on Joining Techniques for Paper Structures*, Dissertation, TU Darmstadt, Institute of Structural Mechanics and Design, 2021.
7 Philip F. Yuan, Xiang Wang, Xiang Wang, "Cellular Cavity Structure and Its Application on a Long-Span Form-Found Shell Design", in: 23rd International Conference of the Association for Computer-Aided Architectural Design Research in Asia, 2018, pp. 297–306.
8 Iasef Md Rian, Dongkuk Chang, Jin-Ho Park, Hyung Uk Ahn, "Pop-Up Technique of Origamic Architecture for Post-Disaster Emergency Shelters", in: *Open House International*, 33(1), 2008, pp. 22–36.
9 Paul Jackson, *Folding Techniques for Designers from Sheet to Form*, London: Laurence King, 2011.
10 Mick Eekhout, Peter van Swieten, *The Delft Prototype Laboratory, Research in Design Series*, 8, Amsterdam: IOS Press, 2015.
11 Mick Eekhout, Tons Verheijen, Ronald Visser, *Cardboard in Architecture*, Amsterdam: IOS Press, 2008.
12 Lawrence C. Bank, Terry D. Gerhardt, "Paperboard Tubes in Structural and Construction Engineering", in: K. A. Harries, B. Sharma (eds.), *Nonconventional and Vernacular Construction Materials. Characterisation, Properties and Applications* (Woodhead Publishing Series in Civil and Structural Engineering), Amsterdam: Elsevier, 2016, pp. 453–480.
13 Matilda McQuaid, *Shigeru Ban*, op. cit.
14 Evgenia Kanli, *Experimental Investigations on Joining Techniques for Paper Structures*, op. cit.
15 Julia Schönwälder, *Cardboard as Building Material*, presentation, TU Delft, 2020, https://docplayer.net/28996964-Cardboard-as-building-material-dipl-ing-julia-schonwalder.html, pp. 47 ff., accessed 10 October 2022.
16 Stefan Schütz, *Von der Faser zum Haus: Das Potential von gefalteten Wabenplatten aus Papierwerkstoffen in ihrer architektonischen Anwendung* (Bauhaus flex research series 1), Weimar: VDG Verlag im Jonas Verlag, 2017.
17 Jerzy F. Latka, "House of Cards – Design and Implementation of a Paper House Prototype", Conference IASS- Interfaces: Architecture, Engineering, Science. Hamburg, 25–28 September 2017.
18 Cottrell & Vermeulen Architecture, "Westborough Cardboard School Building", 2020, https://www.cv-arch.co.uk/westborough-cardboard-building/, accessed 24 March 2021.
19 "Fiction Factory Wikkelhouse", 2020, https://bouwexpo-tinyhousing.almere.nl/fileadmin/user_upload/Wikkelhouse_BouwEXPO.pdf, accessed 24 March 2021. https://www.fictionfactory.nl/en/wikkelhouse, accessed 10 October 2022.

5 LOAD-BEARING STRUCTURE, FIRE PROTECTION, BUILDING PHYSICS

The previous chapter having conveyed the basics of building construction, this chapter deals with concepts for structural dimensioning; aspects of building physics; fire protection; and the possibilities of ecological design, energy balance and recyclability. First, the stability of paper building components is analysed, including constructive aspects related to fire protection. The **»section "Building physics", pp. 75–79**, describes the basic principles of building physics that must be observed for a building envelope made of paper products. The focus is on the hygrothermal properties of paper constructions, i.e. thermal insulation and moisture protection. The life cycle of a paper building is then considered in the framework of sustainability, and the most important parameters for paper constructions suitable for the cycle are presented. Initial findings on the durability of such constructions are also described. Finally, the **»section "Typical construction details", pp. 80–83**, shows in-principle solutions that combine the previously addressed aspects, intending to illustrate the potential functionality of paper products in architecture by way of example.

Dimensioning of load-bearing systems

All components that make up a load-bearing structure (beam, plate, diaphragm, shell and membrane, **»chapter 4, pp. 54–57**) must be dimensioned with regard to their load-bearing functions (stability, serviceability). The dimensioning of a load-bearing structure is an interplay of safety, aesthetics and economic efficiency. To reach an optimal solution, these three aspects must be taken into account in equal measure and reconciled with the boundary conditions set in each case. The solution includes:

- design of the load-bearing structure;
- determination of the impacts;
- modelling, system finding through idealisation;
- internal forces/stresses determination;
- design, dimensioning of building components; and the
- structural design.

In addition, proof of load-bearing capacity and serviceability must generally be provided for every structure to ensure both safety and aesthetics or comfort during use.

Proof of load-bearing capacity

The ultimate limit states are defined to verify load-bearing capacity. This also includes cross-sectional failure. This ultimate limit state is reached when the acting variables reach or exceed the resistance variables. This proof is fulfilled if the following applies:

$E_d \leq R_d$, where:

E_d = rated value of the action variable

R_d = rated value of the stressability

The action variables can be determined from Eurocode 1 "Actions on structures",[1] for example. There is currently no set of rules for building with paper that can be used to determine the design parameters. And there is no recognised procedure for resistance parameters. No standards have yet been set for building with paper either – the closest standards are those dealing with wood as a building material.[2] For timber construction, dimensioning is determined via a characteristic value of the stressability, a modification coefficient and a partial safety coefficient.

$$R_d = k_{mod} \cdot \frac{R_K}{\gamma_M}, \text{ where:}$$

R_K = characteristic strength of a material

k_{mod} = modification coefficient

γ_M = partial safety coefficient

If the so-called characteristic strength of a material is not yet classified or standardised, it can be determined experimentally. Unless specified otherwise, the standard "Basics of Structural Design"[3] specifies a 5% fractile value for common building materials. Accordingly, the characteristic value is the resistivity level below which no more than 5% of the samples fail. The modification coefficient k_{mod} takes into account the duration of the load and the moisture content of the material in the corresponding load situation. In Eurocode 5, which refers to timber construction, the permissible moisture content of the material is specified depending on the use class (i.e. use class A, e.g. residential buildings, and use class B, e.g. garage). Due to paper's hygroscopic and viscous properties, the influence of moisture and load duration must always be considered in relation to timber construction. The partial safety coefficient γ_M takes into account building material properties that can cause an unfavourable deviation from the characteristic value. In the case of paper materials, this allows the recording of influences that stem from damage that may occur in the construction phase during storage and handling or that are caused by different manufacturing qualities. The determination of γ_M and k_{mod} is based on a semi-probabilistic safety concept **» fig. 1**.

» Fig. 1 shows that due to the stochastic distribution of stress and the capacity to withstand stress, there is an overlapping area within which the capacity to withstand stress R is smaller than the stress E. Thus, a component failure is likely, and the verification $E_d \leq R_d$ is not fulfilled. The safety factors of the action and resistance sides must therefore be determined in such a way that the probability of failure (i.e. the overlap area) is minimised in accordance with a required safety level. Empirical values from realised structures and conducted experiments as well as their statistical evaluation are of utmost importance for the semi-probabilistic safety concept. [illegible]

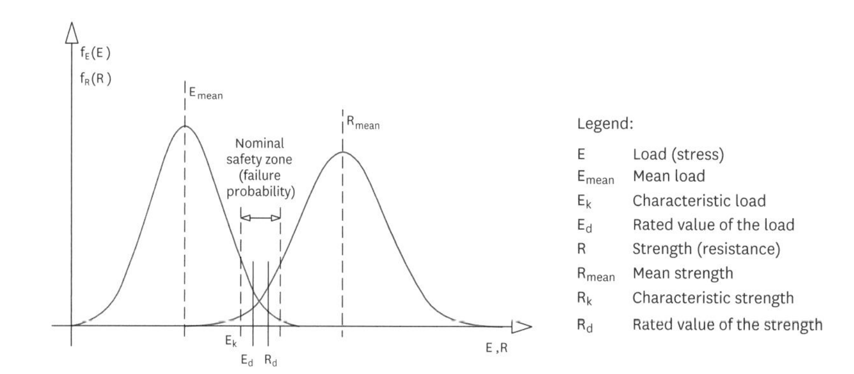

1 Semi-probabilistic safety concept for the assessment of a paper building component.

no meaningful mathematical method of proof based on empirical values, proof is only possible by experiment. These experiments are often carried out based on real-size samples and are therefore correspondingly costly. Calculations should be kept conservative and on the safe side to avoid endangering lives and resources. In addition to the cross-section verifications, the verification of load-bearing capacity includes the verification of position stability, stability verifications, geotechnical verifications and verification regarding the avoidance of fatigue failure of structural components.

Proof of serviceability

In addition to the load-bearing capacity, the serviceability of the structure must also be ensured. This means that the building is not only stable but also (largely) free from vibrations that would make it unsuitable for use. In timber construction, this involves looking at the deformations of the load-bearing structure, which must lie within defined limits. Vibrations of the structure should also be kept within a tolerable range or one that cannot be felt by the users. For bending members with length l, for example, Eurocode 5 specifies the following limiting values of deflections $w_{(net,fin)}$ in the final state:

- $w_{(net,fin)} = \frac{l}{250}$ for beams supported on both sides
- $w_{(net,fin)} = \frac{l}{250}$ for cantilevering beams

The deflection in the final state is made up of the deflections of individual components. The stiffness of the load-bearing structure is adjusted to take into account the influences of moisture and the duration of the load. For this purpose, the deformation coefficient k_{def} is introduced. It allows, for example, determination of a final value of an elastic modulus, E, via $E_{fin} = \frac{E}{1 + k_{def}}$

Here, k_{def} depends on the building material and the use classes mentioned in the » **section "Proof of load-bearing capacity", pp. 69–70**. The elasticity modulus E can be determined experimentally. Due to the high compliance of paper compared with other building materials, it is advisable to define a working range in the force-deformation or stress-strain diagram (a plot representing the deformation due to applied stress). In

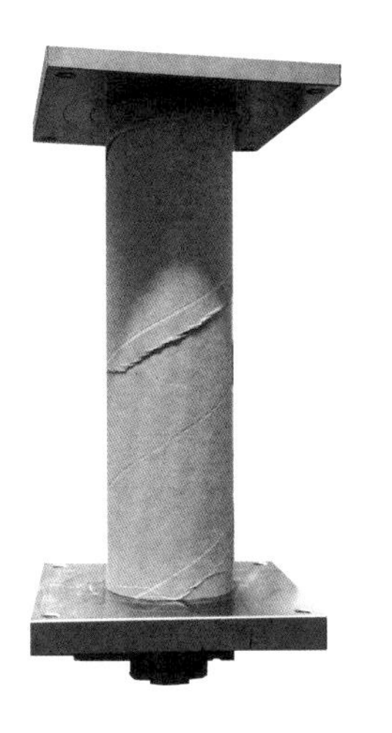

2 Local failure of a tube during an axial compression test, external view (left), internal view (right).

the diagram, the governing elasticity modulus can also be defined as a secant. Deviating elasticity modules can be selected from the working range for proof of the load-bearing capacity.

Structural integrity

In structural calculations, the structural integrity of paper components plays an important role, whereby a distinction is made between local failure on the one hand and global buckling or bulging on the other. (As explained in the previous section, structural integrity is not the same as load-bearing capacity and serviceability; which require further calculations).

Since paper is mainly a thin, flat material, the stability of paper-made components tends to fail under pressure. This can be seen, for example, with axially loaded tubes and paper boards subjected to bending stress, which crease on the side of pressure. This failure is mainly due to the slenderness and imperfections of the material.

» **Fig. 2** shows how such failure occurs locally between the paper web edges. There is clearly more damage on the inside than on the outside. The paper layers bulge inwards in the area of the paper web edge. Another form of stability failure is global buckling or bulging. This failure is structural and occurs with great slenderness, i.e. very tall and thin components, depending on the boundary conditions.

» **Fig. 3** shows force-elongation relationships of edge protection profiles of different lengths as possible structural components under a longitudinally compressive force. The different lengths result in different load capacities and effective stiffness. The following formula can be used to determine the branching load during elastic buckling:[4]

$$F_{\text{krit}} = \pi^2 \cdot \frac{EI}{l_k^2}$$, where:

F_{krit} = branching load

l_k = buckling length

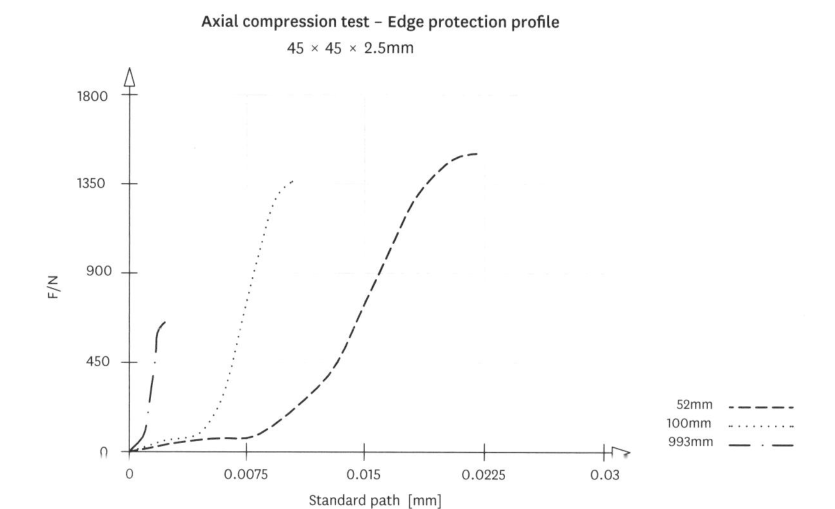

3 Force-elongation diagram of axial compression tests on edge protection profiles (45 × 45 × 2.5mm).

A paper-made support was assumed for the calculation. In one experiment, pressure was applied to the support base with a rigid plate and to the support head with an articulated plate. Under these conditions, Euler buckling case 3 with a buckling length of $l_k = 0{,}669 \cdot l$ can be assumed. With less slender components, the elastic limit can be exceeded before buckling, so the stress-strain diagram leaves the linear range. In this case, the plasticisation of the component must also be taken into consideration. The curves shown in » **fig. 3** refer to components with the slenderness

- $\lambda_{52mm} = 1.7$
- $\lambda_{100mm} = 3.2$
- $\lambda_{993mm} = 31.8$

whereby

$\lambda = \frac{l_k}{i}$, where:

λ = slenderness ratio

i = radius of gyration

and

$i = \sqrt{\frac{I}{A}}$, where:

I = area moment of inertia

A = cross-sectional area

Thus, the 52mm and 100mm long profiles do not have significantly high degrees of slenderness, which is why elastic buckling cannot be assumed in these cases. However, the occurrence of torsional buckling and the influence of imperfections in the assumed paper support are still insufficiently considered and need further investigation.

Fire protection

In paper constructions, structural fire protection is an essential safety concern. If such structures start to burn, it must be ensured that they retain their functions long enough to enable escape and rescue measures as well as extinguishing work. Depending on the building class, the load-bearing structure must remain functional for a correspondingly long period of time. The same applies to components enclosing the spaces. In addition, rapid spread of fire must be prevented. For this reason, no building materials of fire protection class B3 (easily ignited) may be used.

However, since the regulations[5] classify paper as an easily ignited material, it may not be used for construction under building law unless it is combined with other materials so that the flammability reaches class B2 (normal combustibility) or B1 (difficult to ignite). Because paper is still virtually unknown in the building industry, the regulations do not differentiate between different types of it, as is common with other building materials. After all, there are over 3000 different types of paper with different functions and properties. Anyone who wants to use paper in architecture must, therefore, have flammability tests conducted individually according to the current state of knowledge. However, there are some paper materials with B2 certification.

As described in »**chapter 2, section "The flammability of paper", pp. 32–34**, there are various influencing factors that determine the function – in this case, the flammability – of the various paper materials »**fig. 4**. Based on these influencing factors, the materials can be adjusted in favour of structural fire protection via:

- Bulk density and compaction: high compaction and, simultaneously, high bulk density of the material significantly reduce flammability. Compaction reduces porosity and supports the formation of a stable carbon layer in case of fire, which protects the underlying material.
- Surface texture: the denser the surface of the paper, the better the resistance to penetrating flames.
- Fillers and glueing: fillers and the joining of paper materials with glues mixed with, for example, ammonium phosphates significantly improve fire protection.
- Alignment: structured boards, such as corrugated or honeycomb boards in particular, influence fire protection – the decisive factor being the alignment of their cavities. Due to the thermal upwards draft, vertically arranged cavities cause a chimney effect – which accelerates the fire much more than horizontally oriented cavities do.

The addition of fire-retardant chemicals can further reduce the flammability of paper materials. Phosphorus-based additives[6] already protect the material at the fibre level during paper production. The protective effect remains even in the case of a damaged surface. Alternatively, a coating can be applied to protect the surface of the paper from flames and heat »**chapter 2, pp. 33–34**.

Regarding fire protection for paper construction, the construction typologies [illegible]

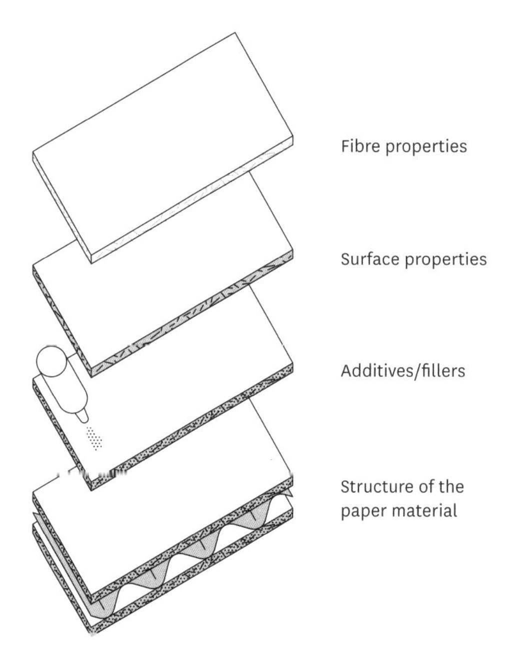

4 Factors influencing the flammability of paper materials. All four material parameters must be considered in the fire protection concept.

two-dimensional structures. Here, the load-bearing structure and outer wall are one unit (relevant for fire protection) and may be enclosed by an insulation layer (without direct fire protection requirements). This is different from the linear structures of frame or skeleton construction, where the load-bearing frame structure is supplemented by room-enclosing elements attached either between or on the outside of it. Sufficient fire protection can be achieved with a layer of compressed solid paper boards, which should meet fire protection class B1 or B2 standards depending on the fire protection requirements. In addition, the construction must also be fire protected on the outside. Since building physics requires the envelope to be water-vapour permeable on the outside, the materials chosen for the outer skin should not exhibit very high bulk density. Fire-retardant additives or water-vapour-permeable, fire-retardant coatings are a better choice.

In the case of lightweight wall construction with corrugated or honeycomb boards forming the load-bearing and insulating structure, it is advisable to add a central solid board layer as a fire barrier in addition to the outer protective levels **»fig. 5**. In this manner, a residual load-bearing capacity is maintained even if the flames have penetrated the outer fire protection layers.

In linear structures, fire protection focuses primarily on the load-bearing, linear construction elements. These must be dimensioned and protected in such a way that, in the event of a fire, they retain their load-bearing function for a period of time corresponding to the building class. These linear elements usually consist of paper tubes **»chapter 3, pp. 42–45**. Depending on the required fire protection, their wall thickness can be adjusted in the winding process during manufacture. In addition, fire-retardant additives can be added to the glues used. The surface of the load-bearing structures can be equipped with a fire-protective coating.

The planar elements of the building envelope have lower fire protection requirements, as they only serve the function of enclosing the space. However, they must

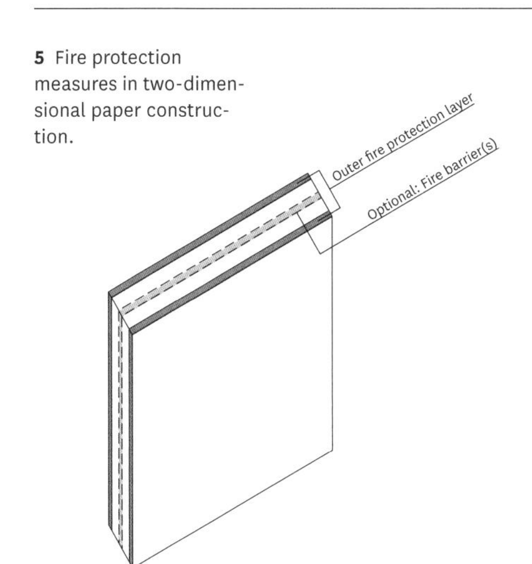

5 Fire protection measures in two-dimensional paper construction.

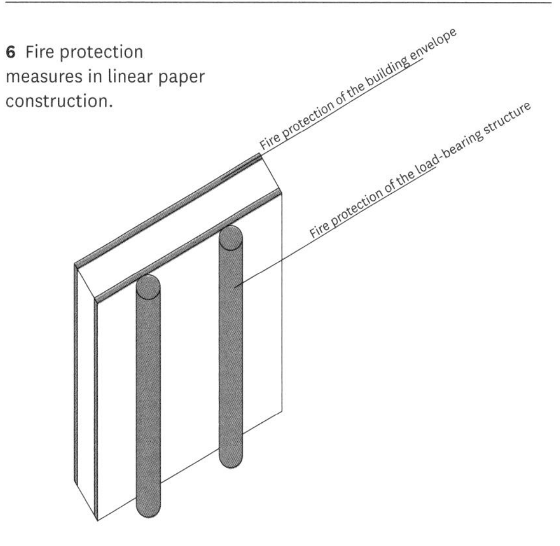

6 Fire protection measures in linear paper construction.

prevent the rapid spread of fire for a certain period of time. Accordingly, linear structures are more economical because only the load-bearing elements need to be equipped and dimensioned for fire protection rather than the entire wall » **fig. 6**.

Building physics

Certain aspects of building physics, such as the thermal and hygric properties of paper materials, play an important role in building with paper. Thermal insulation and moisture protection must be especially carefully planned for the building envelope to ensure a comfortable indoor climate and prevent moisture damage such as mould growth.

Thermal insulation

The better a building is insulated, the less its interior is affected by external temperature changes. The thermal insulating properties of the materials that make up a building envelope are not only essential for a comfortable indoor temperature but also have an impact on heating requirements. Therefore, they significantly influence the energy balance and the associated CO_2 emissions during building operation.

The U-value, i.e. the heat transfer coefficient, is calculated to determine the thermal insulation of a building component. The lower the U-value, the better the thermal insulation. Depending on the installation situation, the U-value is calculated using defined factors, the thickness and the thermal conductivity of the materials used.

» **Fig. 7** lists averaged thermal conductivity values λ_{mean} of different paper materials based on material tests. They are determined using a hot plate device for measuring » **chapter 8, p. 178**. These values provide initial estimates of the thermal insulation properties of paper structures.

7 Thermal insulation properties of paper materials, the necessary component thickness for a U-value of 0.21 W/(m² · K) and the required material quantity.

	Bulk density [kg/m³]	λ Mean [W/(m² · K)]	Component thickness* [m]	Material requirement** [kg]
Cellulose flock	30–80	0.04–0.045	0.18–0.21	5.4–16.8
Corrugated board	130	0.051	0.23	29.9
Honeycomb board	50	0.065	0.3	15
Corrugated multi-wall board	50–100	0.072–0.09	0.33–0.41	16.5–41
Solid board	450–475	0.07–0.072	0.32–0.33	144–156.75

* Calculated component thickness to achieve a U-value of 0.21 W/(m² · K)
** Material requirement related to 1m² wall surface

8 Different temperature gradients occur in a component depending on the orientation of the cavities: a. corrugated board, orientation of the cavities orthogonal to the temperature gradient, b. honeycomb board, orientation of the cavities parallel to the temperature gradient.

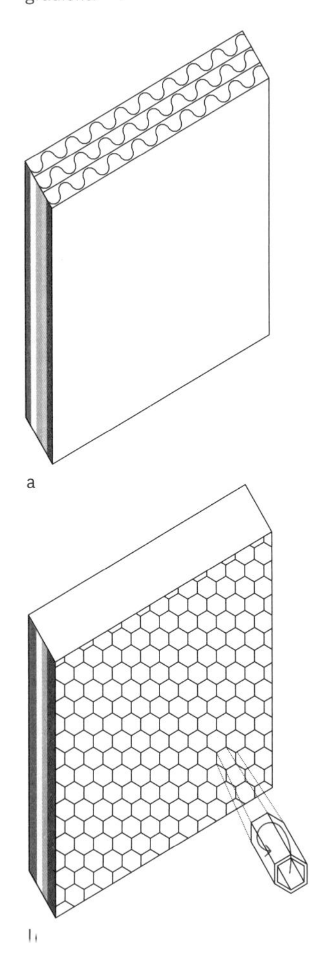

The thermal conductivity of paper materials depends on their bulk density and structure. Cellulose flock is already established in the building industry as blow-in insulation. It has the lowest thermal conductivity due to its low bulk density and flock structure, and thus scores with the best insulation values. Corrugated board has a very good insulating effect despite its higher bulk density due to its enclosed air spaces. If the cavity runs parallel to the temperature gradient, for example, in the case of corrugated multi-wall and honeycomb boards, convection and thus increased thermal conductivity can occur depending on the circumference and depth of the cavity **»fig. 8**.

In order to better understand how the different thermal conductivities affect each other, component thickness and the material requirements for manufacturing such components are derived from the specific material parameters. However, this approach is purely theoretical as the materials listed can only be used in material combinations, for example with solid board, due to the moisture and fire protection requirements explained below.

Components made of corrugated board and cellulose flock can feature the lowest component thickness due to their low thermal conductivity. Although the thermal conductivity of corrugated multi-wall board and honeycomb board is higher, they are also suitable as insulation materials as the material requirements are still within limits. Insulation with solid board, on the other hand, is not recommended as the necessary component thickness and material requirements are unreasonably high.

Compared with conventional building materials, common paper materials perform quite well in terms of their insulating properties. Although they do not reach the thermal conductivity of classic insulation materials such as EPS, XPS (expanded and extruded polystyrenes, respectively) and PU with values between 0.03 W/(m² · K) and 0.035 W/(m² · K), they are comparable with wood wool and wood fibre boards and insulate much better than mineral or metallic building materials.

The insulation capacity of a paper structure also depends significantly on the connection techniques described in »**chapter 4, pp. 57–62**. Local thermal bridges can occur depending on the type of connection and the material used »**fig. 9**. Mechanical joints made with highly thermally conductive materials such as metals – screw connections, nails or rivets – are inevitably accompanied by thermal bridges. In the worst case, such connections lead to condensation. The following section explains how condensation occurs and what preventive measures can be taken. Moisture is particularly problematic in the area of the joints because the paper structure dissolves when exposed to water unless it has been modified. If that happens, the fixation no longer has a hold and can fail.

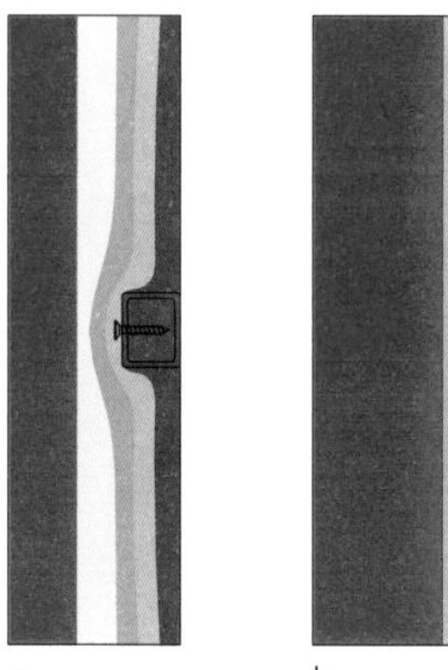

9 Thermal bridges resulting from joining techniques that conduct thermal energy better than the material itself due to different material properties (for example, a. mechanical fixing, or b. adhesive joints).

Moisture protection

As already described in »**chapter 2, pp. 28–32**, the moisture content of a paper material has a significant influence on its mechanical properties because water dissolves the hydrogen bonds of the cellulose fibres in the paper. However, drying can largely restore these bonds. Therefore, paper is a capillary-active material, which means that paper can absorb, transport and release water.

Moisture not only affects the mechanical properties of a paper component but is also considered the main cause of mould growth and thus an unhygienic indoor climate. In the building industry, capillary-active materials are often used in moisture-critical areas to prevent mould growth – one typical example being interior insulation. In this respect, paper materials have significant advantages over other building materials.

To prevent mould from developing, DIN 4108-3 specifies that the amount of condensation that occurs in a building component during the winter period may not be larger than the amount that can evaporate again in the summer.[7] However, since condensation damages the structure of paper, paper-made buildings must be designed in such a way that no condensation can occur.

There are several structural approaches to achieving this goal. On the one hand, the components can be constructed in such a way that the layers facing the interior have a significantly greater water vapour diffusion resistance than the outer layers. This ensures that water vapour can diffuse out of the component and does not condense within the component and thus form condensation. The water vapour diffusion resistance is considerably greater in highly compressed cardboard with low porosity than in paper materials with low bulk density and high porosity. Thus, highly compressed cardboard should be used for the inner layers and more porous paper materials for the outer layers of a component. Another solution could be ventilated façades. These facilitate the evaporation of water vapour through continuous air circulation, and are therefore strongly recommended for paper constructions.

The water vapour diffusion density is also significantly influenced by the connection technology. Mechanical fixings or penetrations of any kind can form weak points and thus contribute to condensation. However, they can also be used for targeted water conduction in the component. Adhesive joints can increase or even reduce the water vapour diffusion density – depending on the glue chosen and whether it is applied over the entire surface or at specific points.[8]

The water vapour diffusion resistance factors of paper materials in »**fig. 10** show a wide range. They are lowest in multi-wall corrugated board and regular corrugated cardboard due to the high air content and open-pored structure. However, the situation

10 Water vapour diffusion resistance factors of different paper materials.

	Bulk density kg/m³	Water vapour diffusion resistance factor [μ]
Corrugated multi-wall board	50–100	2–5
Corrugated board	130	11–24
Solid board not compressed	390–760	16–54
Solid board compressed	750–780	97–107
Cardboard in laboratory environments	740–770	151–177

is entirely different with compressed solid boards. Lab-produced cardboards with even lower porosity than normal have an even higher density. This allows complete external wall structures made of paper that meet the requirements for moisture protection without condensation occurring in the construction; however, a defect-free design is a prerequisite.

The water vapour diffusion resistance factors of paper materials are comparable with those of many proven building materials – such as wood-based materials, insulation materials and mineral materials:

- Wood fibre boards 3–30μ
- Plywood boards 50–250μ
- Mineral wool insulation 1μ
- XPS insulation material 80–250μ
- Mineral plaster 10–35μ
- Concrete 60–130μ

The values of plastics, on the other hand, are much higher (5000–100,000μ). Glass and metal materials are entirely impermeable to water vapour, which is why they are specified with ∞ μ.

Paper constructions must be protected not only from water vapour but also from direct water penetration caused by rain or splash water. There are two possibilities to achieve this: additives are added to the paper pulp so that water can penetrate but does not loosen the fibre bonds. Or a subsequent coating such as a film is applied to protect the paper surface and prevent water from penetrating. However, both measures can have a negative impact on the recyclability of the paper materials. Therefore, each individual case must be carefully assessed to determine which procedure makes sense. Additional constructive moisture protection can be implemented, for example, by roof overhangs, elevated construction methods and double façades »fig. 11.

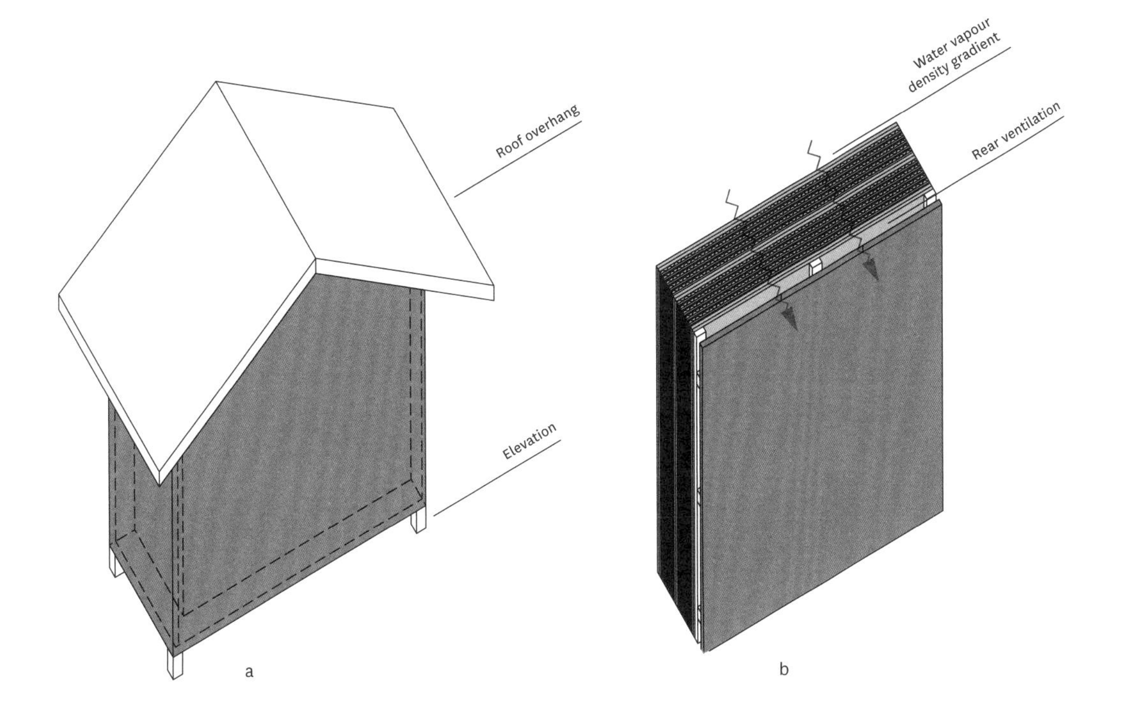

11 Moisture protection:
a. structural moisture protection measures through roof overhang and elevated buildings and
b. protection against condensation through appropriate insulation thickness.

Ecological considerations

Paper consists of up to 100% of the renewable raw material wood. As a building material, paper consequently expands the range of building materials based on renewable raw materials and thus serves climate and environmental protection goals, resource conservation and CO_2 reduction.

Wood waste such as wood shavings from sawmills, thinning wood, crooked trees and trees with defects or many knotholes are also suitable as raw material. Thus, paper production does not require high-quality trees as construction timber does. During the manufacturing processes, paper materials pose a primary energy demand (non-renewable) of 1.77 to 10.83 MJ (Megajoule) and CO_2 equivalent emissions of -1.65 to -0.33 kg CO_2 eq. The CO_2 emissions show negative values because trees bind CO_2 in the wood during their growth phase, which remains in the paper fibres and is only released through combustion or decomposition at the end of the life cycle. The ecological parameters of paper production are therefore comparable with those of wood-based materials and can be assessed to be similarly positive in terms of environmental and climate protection.

The recyclability of the material is another ecological and economic advantage. Paper recycling processes have been around for several centuries and have been greatly optimised, especially in recent decades. Thus, paper features a well-functioning and closed material cycle. For paper constructions, however, the paper materials must be removed from this closed loop and run through multiple, longer recycling cycles with component production, building construction, building use and maintenance, demolition and their return to the paper cycle → fig. 12.

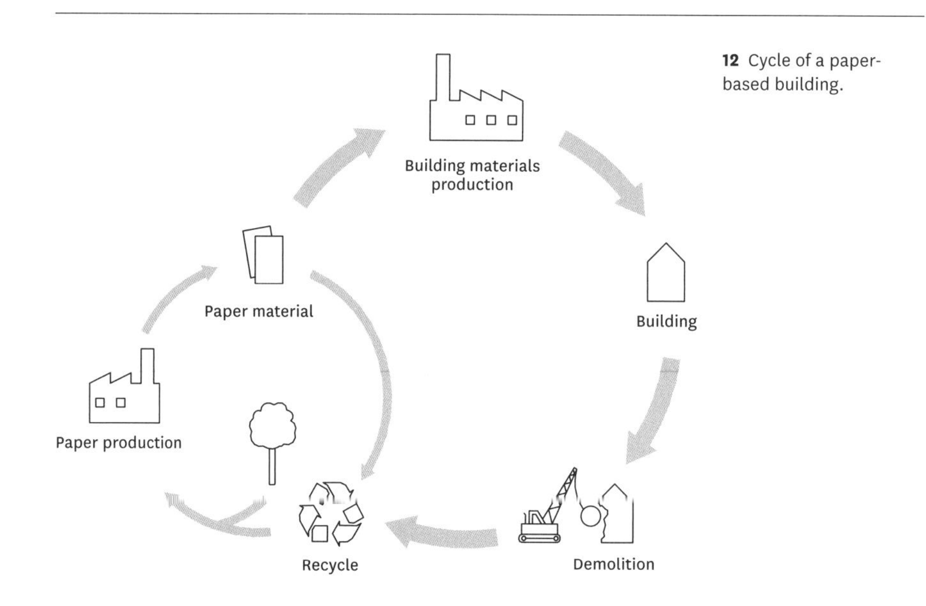

12 Cycle of a paper-based building.

In order to return the paper components to the paper cycle, their materials must be separable from other materials and also from each other. Ideally, the separation will have already taken place during selective dismantling. So, even during the planning phase, the connection techniques within the components must be planned reversibly, i.e. they should be able to be detached non-destructively. If this is not possible, separating the different materials must occur in the recycling process. From an ecological point of view, mechanical fixings (e.g. screws) or flexible connections (e.g. seams) are preferable to adhesive connections, which are difficult to separate.

The durability of a paper structure strongly depends on the previously described mechanical building physics properties of its paper construction. The more resistant the construction is to water and mechanical stress, the longer it can be used.

Since building with paper does not have a long history, there is a lack of long-term experience with paper structures. Only pilot projects like those described in **»chapter 6** offer initial experience. Most of these projects are or were designed as temporary solutions, such as the Japanese Pavilion (Expo 2000) **»pp. 132–133**, the Cardboard Theatre Apeldoorn **»pp. 130–131**, and the Paper Theatre in IJburg **»pp. 134–135**. However, there are also durable paper structures, such as the Westborough School's cardboard building **»pp. 96–99**, which was designed to last 20 years. Fiction Factory guarantees a durability of 15 years for the Wikkelhouse **»pp. 90–95**, but estimates a usability of 50 years.

Typical construction details

The following section presents in-principle solutions for building with paper, based on the systematics of beams, plates, diaphragms, shells and membranes. These solutions include suitable options for load-bearing structures, foundation and roof components. This section also contains descriptions of possible techniques of joining paper materials with each other and with connecting components.

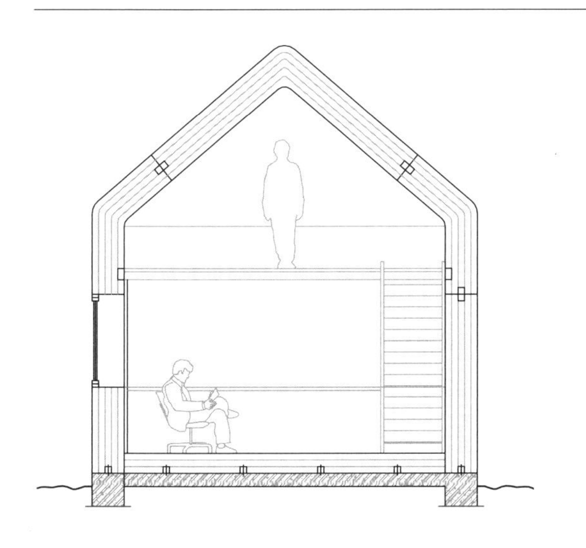

13 Vertical section of a plate (diaphragm) structure, scale 1:200. Representation of load-bearing walls made of several layers of paper and corrugated board.

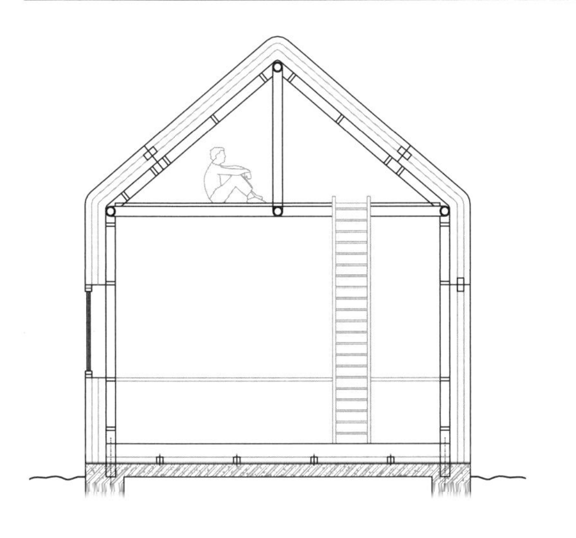

14 Vertical section of frame construction, scale 1:200. Representation of the inner load-bearing skeleton structure and the enveloping façade, both made of paper.

15 Roof detail of a plate (diaphragm) structure, vertical section, scale 1:40. The plate structure is made of several paper layers and the window connection is shown.

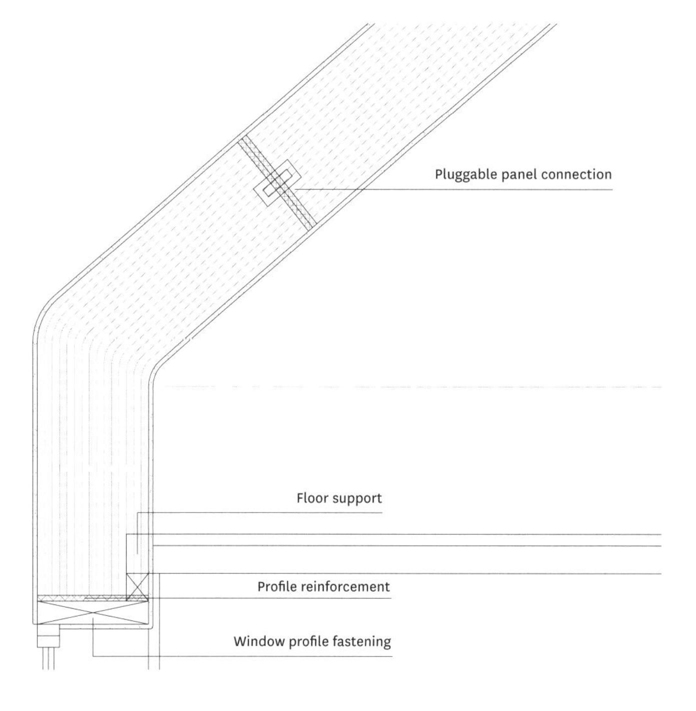

17 Façade detail of a plate (diaphragm) structure, horizontal section, scale 1:40.

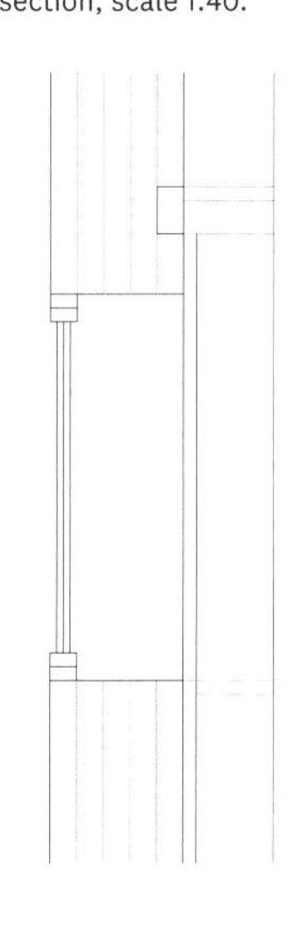

18 Façade detail of a frame structure, horizontal section, scale 1:40.

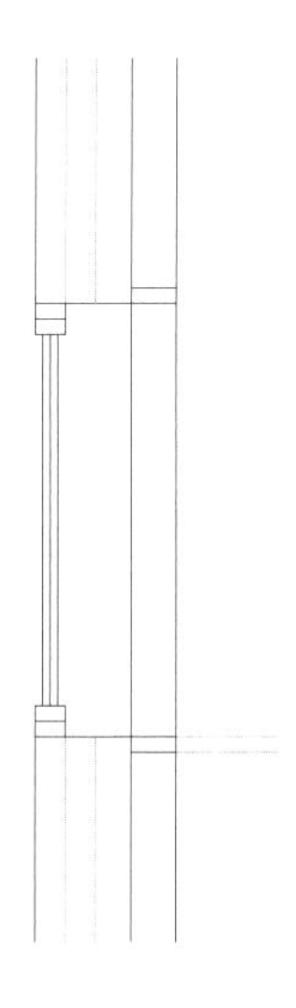

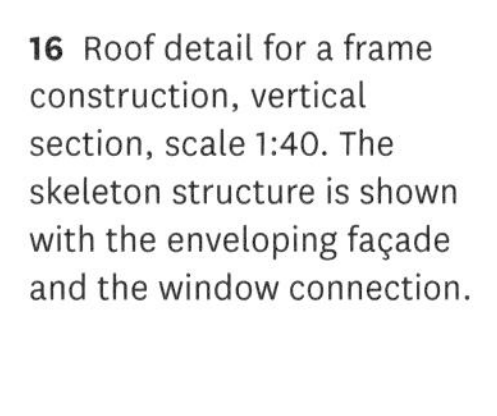

16 Roof detail for a frame construction, vertical section, scale 1:40. The skeleton structure is shown with the enveloping façade and the window connection.

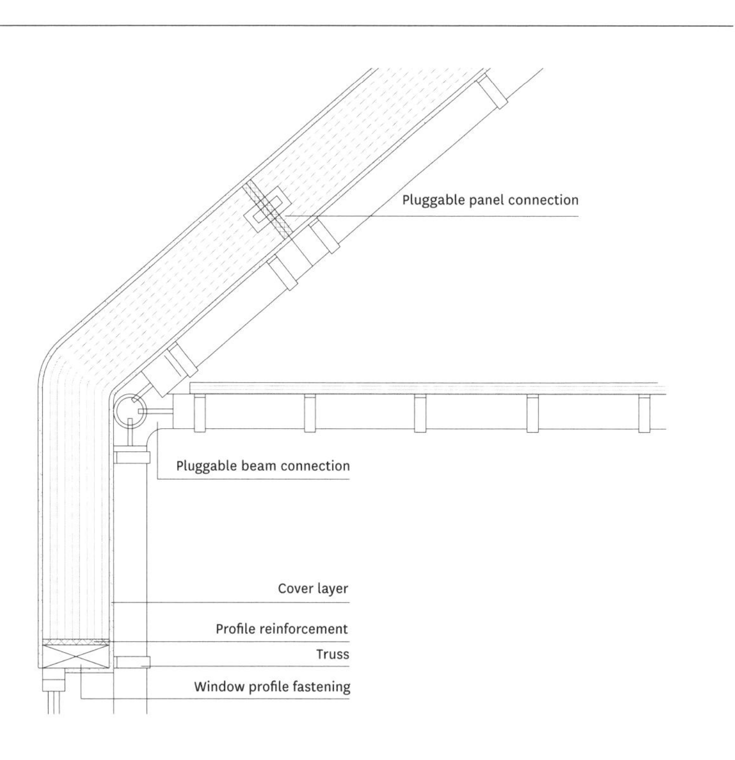

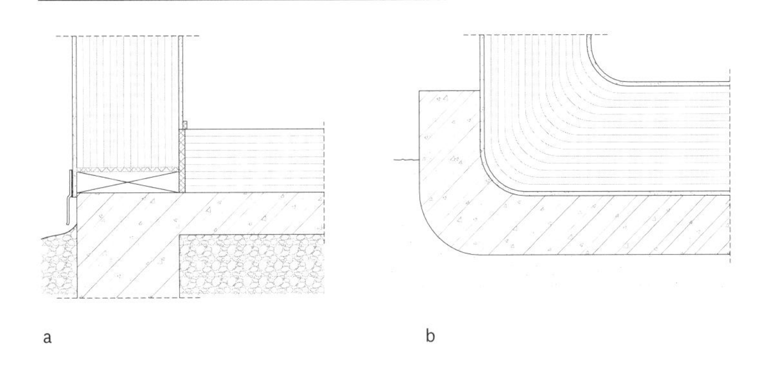

a b

19 Concrete foundation of a plate (diaphragm) structure:
a. load transfer from the wall directly into the foundation,
b. interior floating paper floor.

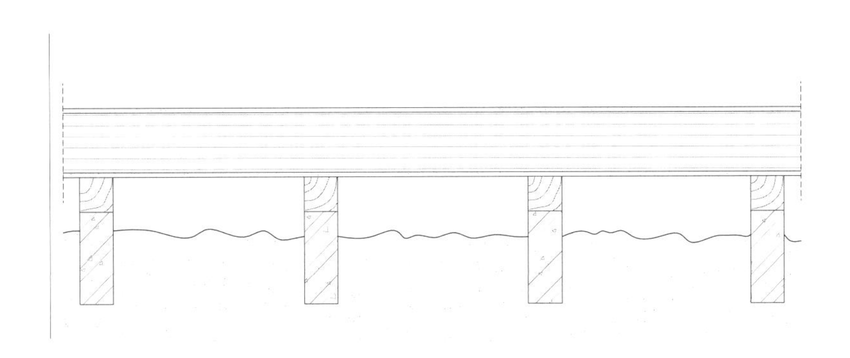

20 Strip concrete foundation with load-bearing wooden beams and paper floor.

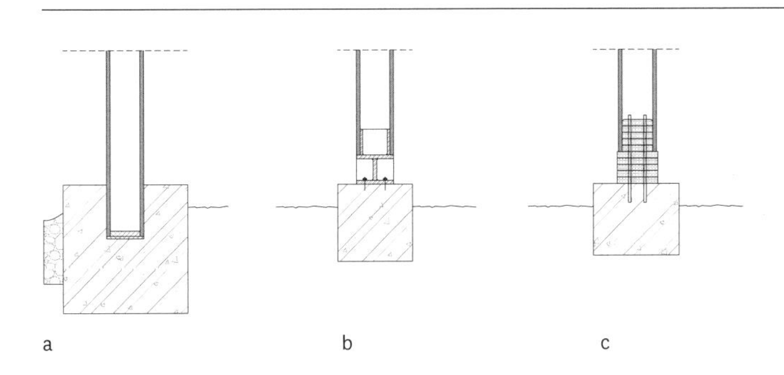

a b c

21 Examples of foundation solutions for cardboard sleeves:
a. Concrete,
b. steel,
c. timber.

REFERENCES

1 *DIN EN 1991-1-1. Eurocode 1: Actions on Structures – Part 1-1: General Actions – Densities, Self-Weight, Imposed Loads for Buildings*. Berlin: Beuth, 2010.
2 *DIN EN 1995-1-1. Eurocode 5: Design of Timber Structures – Part 1-1: General – Common Rules and Rules for Buildings*. Berlin: Beuth, 2010.
3 *DIN EN 1990. Eurocode: Basics of Structural Design*. Berlin: Beuth, 2010.
4 Dietmar Gross et al., *Technische Mechanik 2: Elastostatik*, Berlin/Heidelberg: Springer Verlag, 2012.
5 Deutsches Institut für Bautechnik (n.d.), Informationsportal Bauprodukte und Bauarten, https://www.dibt.de/de/bauprodukte/informationsportal-bauprodukte-und-bauarten, accessed 7 August 2020.
6 Maryam Ghanadpour, Federico Carosio, Per Tomas Larsson, Lars Wagberg, "Phosphorylated Cellulose Nanofibril: A Renewable Nanomaterial for the Preparation of Intrinsically Flame-Retardant Materials", in: *Biomacromolecules*, (16)10, 2015, pp. 3399–3410.
7 *DIN 4108-3:2018-10 Thermal Protection and Energy Economy in Buildings – Part 3: Protection Against Moisture Subject to Climate Conditions – Requirements, Calculation Methods and Directions for Planning and Construction*, Berlin: Beuth, 2010.
8 Rebecca Bach, *Papier Fassaden – Entwicklung konstruktiver Prinzipien für Fassaden aus Papierwerkstoffen mit Fokus auf Brandschutz, Wärmedämmung, Feuchteschutz und ökologische Eigenschaften*, Dissertation, RWTH Aachen, 2020.

6 CASE STUDIES

Today's increasing demand for small, portable, mobile and temporary structures is closely linked to our current volatile living conditions, described as the era of "liquid modernity" – a term coined by Polish-British sociologist and philosopher Zygmunt Bauman. Liquid modernity means that we are less and less connected to a particular place. Among the causes are social networking activities that take us virtually anywhere on the globe. Traditional spatial and social references have given way to a sequence of new beginnings. We prefer to pop up here and there for a short while rather than stay in one place for a longer period of time. The pop-up phenomenon creates the need to generate spaces for temporary use. These include pop-up restaurants or pop-up stores, offering new opportunities for neighbourhoods and small entrepreneurs. Paper structures such as the Japanese Pavilion for the Expo in Hanover in 2000 or the Apeldoorn Theatre in 1992 considerably influenced such pop-up constructions.

But temporary use also includes emergency shelters after earthquakes, floods or military conflicts. In this context, building types can be distinguished depending on the intended period of use. Emergency shelters are designed to be used for a few weeks and can provide refuge for many people at once. This type of structure includes projects such as the Paper Log House »**pp. 88–89**, TECH 04 »**pp. 118–119** or the Paper Emergency Shelters »**pp. 126–129**.

Yet another reason for the trend is urbanites' longing for quiet retreats. It stimulates the demand for small weekend houses that offer very modest infrastructure. Examples are the Paper House »**pp. 86–87**, a weekend get-away by Shigeru Ban, and the Wikkelhouse »**pp. 90–95**. Both holiday homes were created from paper-based components without compromising on amenities.

In addition, there is the Tiny House movement, which generally propagates living in a small space. The agenda here is a reduced and resource-conserving lifestyle. The Cardboard House in Sydney »**pp. 100–101** or the House of Cards »**pp. 114–117** represent such low-cost, tiny living spaces.

The built examples and projects in this chapter are divided into four groups. The first group includes houses and shelters »**pp. 86–129**, i.e. paper structures, that are closed on all four sides. They are designed to provide shelter from the outdoor climate.

Another type of structure made of paper materials is pavilions »**pp. 130–137**. In contrast to houses and shelters, pavilions form spatial structures with no enclosed building envelope. They serve as spatial boundaries for temporary events. In addition, pavilions can provide sun protection, such as Cardboard Theatre Apeldoorn »**pp. 130–131**, and shield off rain, as the Japanese Pavilion at Expo 2000 in Hanover »**pp. 132–133**.

The third group presents temporary bridge constructions made of paper materials »**pp. 138–143**; the PaperBridge is made entirely of layered copy paper »**pp. 140–141**,

Shigeru Ban's and Octatube's Pont du Gard bridge **»pp. 138–139** was constructed mainly of cardboard tubes, and the IPBU Bridge in Darmstadt **»pp. 142–143** was built of various paper materials.

The chapter concludes with interior furnishings and furniture made of paper materials **»pp. 144–157**. While the Aesop DTLA store **»pp. 144–147** is entirely made of cardboard tubes, the Cardboard Office in Pune uses partitions made of honeycomb cardboard panels **»pp. 152–155**. Apart from the classics of the Carta Collection **»pp. 156–157**, the furniture by Stange Design and the interiors by Nordwerk are interesting examples of cardboard applications. There are already many designer pieces available that are intended to be affordable for the mass market.

PAPER HOUSE

ARCHITECT/INVENTOR: Shigeru Ban
LOCATION: Lake Yamanaka, Yamanashi, Japan
YEAR: 1995
USE: Living
CONSTRUCTION TYPOLOGY: Column construction made of cardboard tubes
AREA: 100m²
PLANNED SERVICE LIFE: Permanent

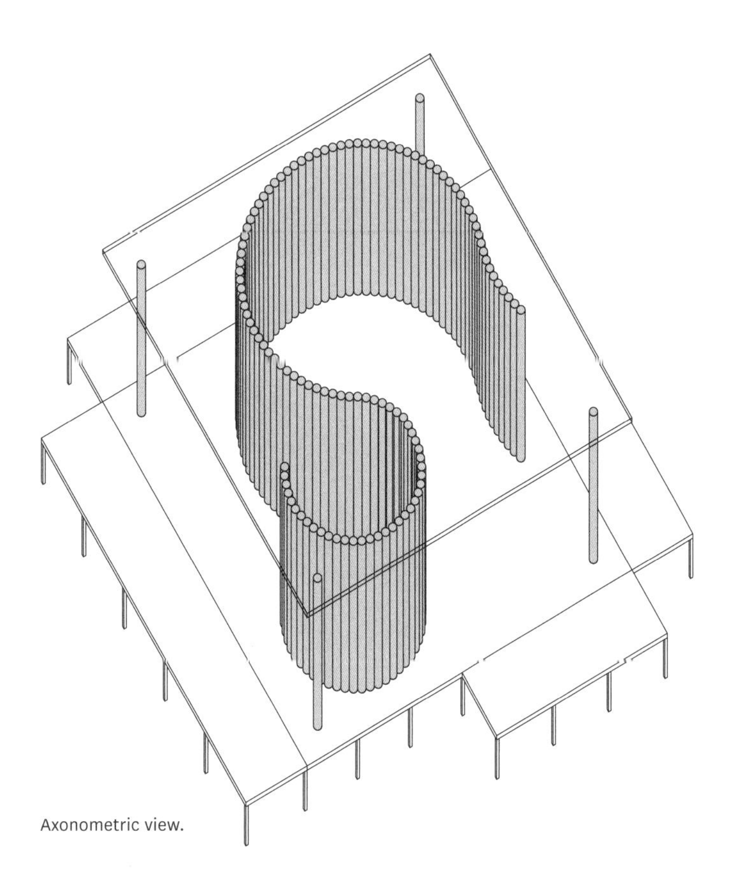

Axonometric view.

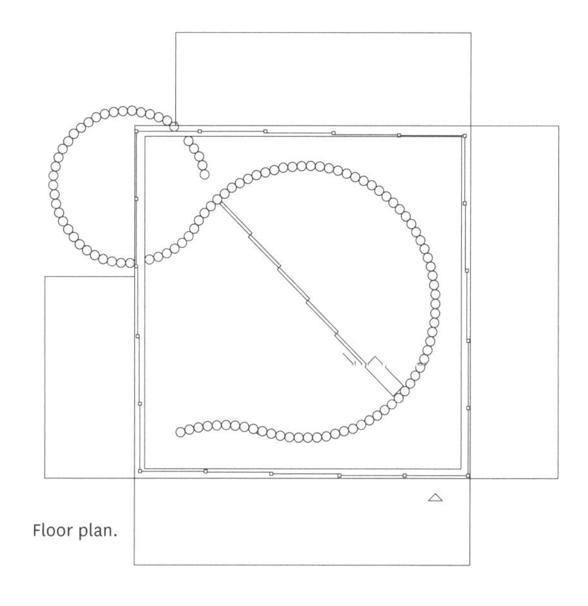

Floor plan.

Paper House by Shigeru Ban was the first project to receive approval for a permanent structure made of cardboard tubes. The building, which Ban himself uses as a weekend home, stands on a sloping site near Lake Yamanaka in Yamanashi Prefecture. The 10 × 10m house consists of a total of 110 cardboard tubes arranged in an S-shaped pattern. They separate the house's interior from the small garden next to the bathroom. Seventy-five of these cardboard tubes form a large circle that, in turn, separates the living area from the surrounding gallery. The living area is unfurnished except for a free-standing kitchen counter and movable cupboards. The cardboard tube walls frame the view of the nearby forest and thus create a remarkable continuity between inside and outside. The corridor surrounding the living area leads to a free-standing 127cm diameter cardboard column that hides a toilet. From the prominent entrance to the house, the row of columns leads to a small cardboard tube circle bordering a bathroom with a small garden. The living space is framed by a semi-circle of cardboard tubes that serves as a privacy screen. The small gaps between the cardboard tubes provide the interior with daylight, while in the evening the house glows discreetly

Exterior view.

The square area of the house is enveloped by sliding glazing, which refers to traditional Shoji panels and allows the occupants to directly access the rectangular terraces outdoors. The house can thus be opened up completely and can literally merge with the surroundings, with its white roof line standing out visually.

View from the interior.

PAPER LOG HOUSE

ARCHITECT/INVENTOR: Shigeru Ban
LOCATION: Kobe, Japan
YEAR: 1995
USE: Temporary accommodation
CONSTRUCTION TYPOLOGY: Column construction made of cardboard tubes
AREA: 16m²
PLANNED SERVICE LIFE: Temporary

Axonometric view.

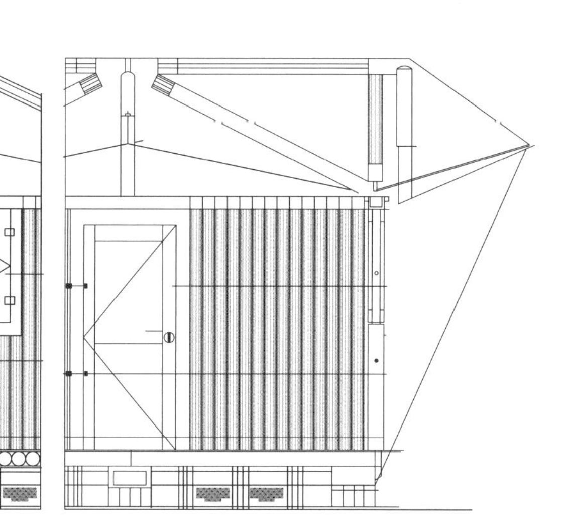

Section-elevation.

The Great Hanshin Earthquake, measuring 7.3 on the Richter scale, hit the city of Kobe in January 1995. The natural disaster made 320,000 people homeless overnight, and over 240,000 houses were damaged. Those who could not find accommodation with relatives or acquaintances had to seek emergency shelters; ethnic minorities inhabited the local parks. Shigeru Ban proposed to build small housing units for the Vietnamese migrants staying in the parks in Kobe's Nagata district. The Paper Log Houses were made in a nearby workshop from materials available on site.
Each of the houses covered an area of 6 × 6m and could accommodate one family. The building construction consisted of a foundation, floor elements, paper tube walls and a textile PVC roof.
Beer crates were filled with sandbags for the foundation. A wooden floor slab was laid on top of the crates. The wall panels, each measuring 2×4m, consisted of 2m high cardboard tubes (ø 108mm) with a wall thickness of 4mm. The tubes were mounted next to each other and taped

Paper Log House at the exhibition in Mito, Japan, 2013.

Interior view.

with sealing tape for better thermal insulation. First, the tenons of the 4m long and 12mm thick plywood panel were pushed onto the cardboard tubes from below; then, the component was mounted and screwed to the wooden floor panel.

In addition, 6mm steel bars ran horizontally along one third and two thirds of the wall height – they provided additional stability and enabled the installation of window and door frames. At the top, the walls were closed off with plywood connectors. Wooden beams and perforated L-profiles formed the transition to the roof. The roof frame was mounted on the wooden girders. The frame was made up of cardboard tubes that were fastened together with plywood joints. A membrane was stretched over the frame, attached to the perforated L-profiles and additionally anchored to the foundation with ropes. The gable roof was equipped with an additional insulation layer inside the house. The gable walls of the roof could be opened for ventilation. After assembly, the paper tubes were painted with solvent-based varnish to protect them from rain and moisture.

The houses were grouped next to each other. The 1.8m wide spaces between them were each shared by two families as a sheltered outdoor area. In 1995, a total of 27 Paper Log Houses were built for both Vietnamese and Japanese occupants. They were in use for four years.

The concept of the Paper Log House proved itself several times in emergency situations; of course, adapted to the local conditions. In 2000, 17 Paper Log Houses were built after an earthquake in Turkey. Due to the typically larger family size, these houses had a larger floor area, and the cardboard tubes were filled with waste paper for better insulation.

One year later, 20 houses were built in India. As there were not enough beer crates available in India, the foundation was built from the rubble of buildings destroyed by an earthquake and covered with a flooring. The Indian version had a domed roof made of cane mats with a tarpaulin laid on bamboo ribs. A shaded veranda served as a space for daily activities.

In 2014, after an earthquake in the Philippines, Ban created the fourth version of the Paper Log House. This time, the cardboard tubes were only used for the frame construction, covered with locally produced air- and light-permeable bamboo mats.

WIKKELHOUSE

ARCHITECT/INVENTOR: René Snel;
further development: Fiction Factory
LOCATION: not specified,
production in Amsterdam
YEAR: 1996 (invention);
further development since 2012
USE: Living
CONSTRUCTION TYPOLOGY:
Frame construction
AREA: variable, 5m² per segment
PLANNED SERVICE LIFE: 50 years
(expected), with 15-year warranty

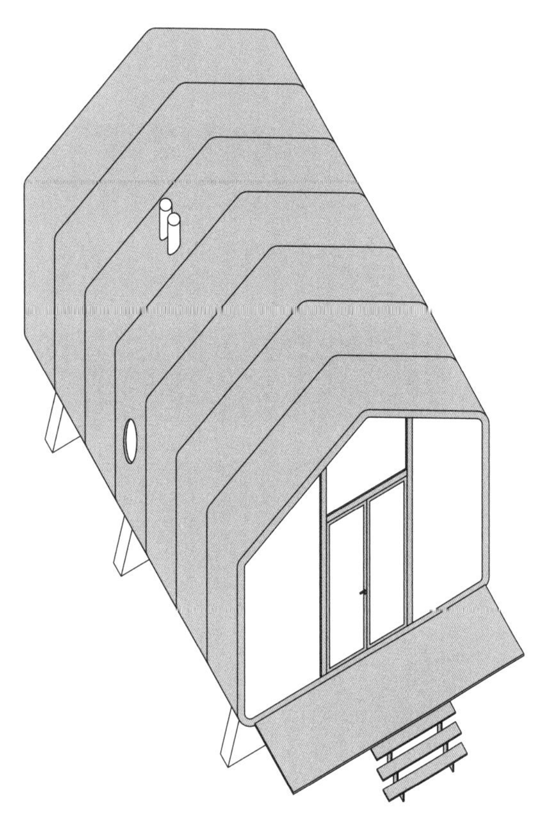

Schematic exterior view.

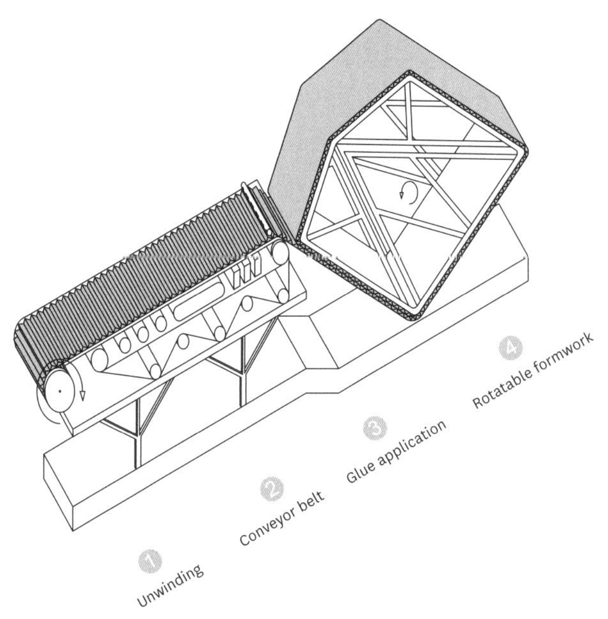

Production process of the Wikkelhouse: the self-supporting building envelope is created by wrapping single-layer corrugated cardboard around a house-shaped formwork.

The Wikkelhouse was one of the first mass-produced paper-based residential buildings. René Snel's concept is based on a machine he invented that wraps and glues several layers of corrugated cardboard around a house-shaped formwork. Several of the resulting elements can then be attached to each other to form a small house.
Originally intended as an emergency shelter for disaster areas, Wikkelhouse was to be easily and quickly transportable. Therefore, Snel designed the winding machine to fit on a truck for efficient transport, at least over short distances. The source material could also be transported by truck in its most compact form, namely rolled up. The houses could then be manufactured directly on site and according to individual requirements.
However, the demand was so limited that Snel sold the Wikkelhouse idea and machine to the Fiction Factory company in 2012. Since then, Fiction Factory has continued to develop and market the Wikkelhouse concept at its production site in Amsterdam for permanent residential use. In practice, this means that the house is mainly used as a holiday home. Originally mounted on a truck, the winding machine now stands in a production hall, enabling higher production rates and fast prefabrication of the individual elements.
The high degree of prefabrication in

Exterior view.

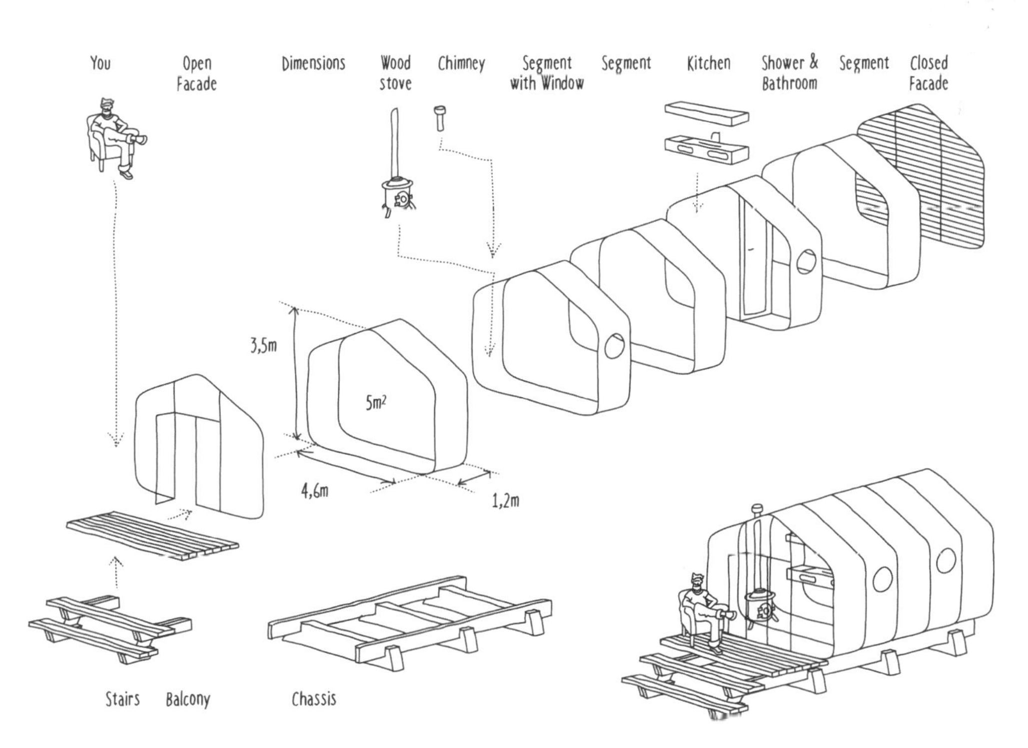

[illegible]

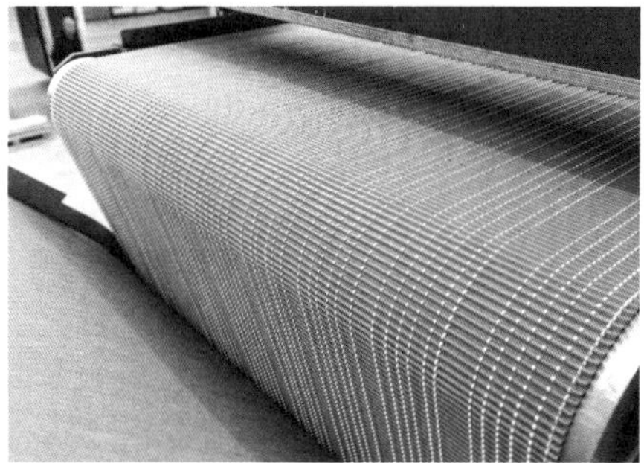

Wikkelhouse machine, glueing machine, module production in the production hall.

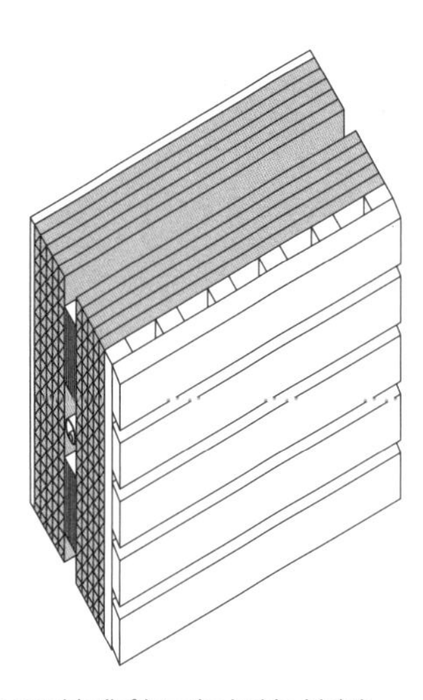

Detail of the wall construction of the Wikkelhouse: from the outside (left) to the inside (right), the wall consists of wooden battens, waterproof foil, 12 layers of corrugated cardboard, an installation layer with fasteners, another 12 layers of corrugated cardboard and wood veneer.

characteristic for the Wikkelhouse. It is produced and sold in individual segments (width/height/depth: 4.6/3.5/1.2m). Different segment typologies can be chosen and arranged individually to design a custom living space. There are segments with integrated windows; segments with a wet cell for the bathroom; and various options for installing building services, such as a fireplace for heating and kitchen elements. The enclosures on the gable ends are either transparent glass or opaque façades. Access is usually gained through a recessed front door in the transparent glass front.

The highlight of the Wikkelhouse is its production technique: as the Dutch word *wikkelen* (to wrap, to wind) suggests, the elements are produced by wrapping single-wall corrugated cardboard around a house-shaped formwork. The corrugated cardboard is wrapped around the formwork in 24 layers, creating an approx. 20cm thick wall core that features load-bearing as well as insulating properties. The U-value of this construction, i.e. the coefficient of heat transfer of a building component, is approx. 0.24 W/(m²·K). Compared with other construction designs, it does not yet meet the passive house standard, but it can be classified as energy-efficient.

A layer of veneer wood is usually applied on the inside, but alternative interior surfaces can also be provided on request

Installation layer in the wall structure.

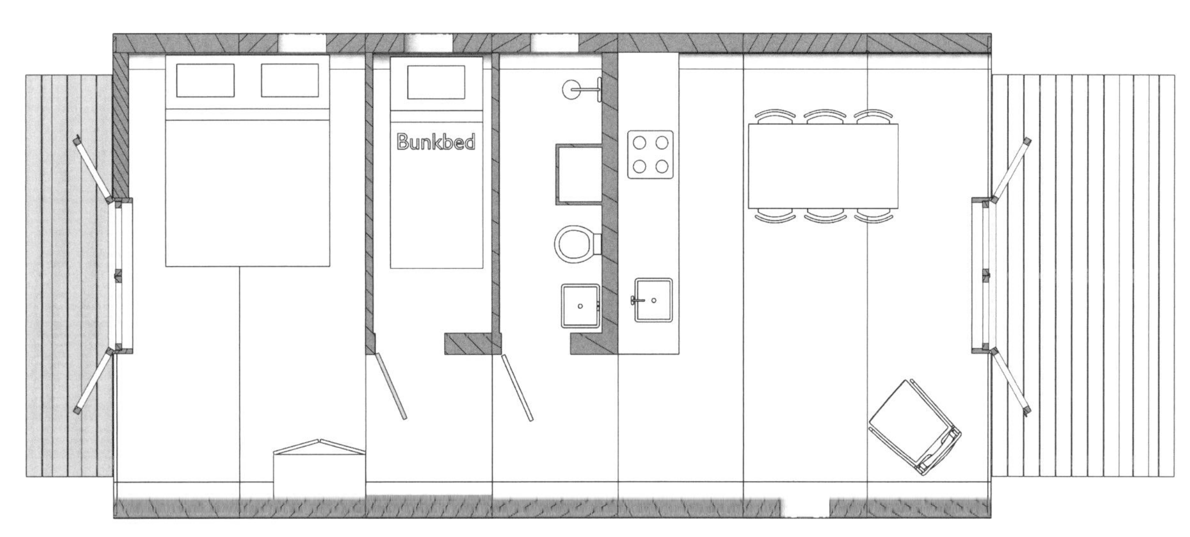

Floor plan of a potential subdivision

The Wikkelhouse is manufactured in segments and assembled on site (top). It can also function as a floating holiday home, shown here during transport and at its anchorage in the port of Rotterdam (bottom).

On the outside, the wall core is covered with a waterproof yet diffusion-open foil and mechanically fastened.
The foil prevents water from penetrating the structure from the outside but allows moisture to diffuse out of the component. Therefore, mould formation and structural weakening due to moisture in the wall core are avoided. The outermost layer is a rear-ventilated façade, usually made of horizontal timber battens. These shield the construction from rain, splash water, etc. The air circulating behind it additionally protects the construction from moisture.
Wooden elements at the edges of the segments enable connection between the segments. These wooden connectors are located identically in each segment so that the different segment variants can be freely combined.
Including all structural elements, the wall construction reaches a thickness of about 25cm. Due to the construction method of winding and glueing the entire house in one process, the roof and the base construction are analogous to the construction set-up described above.
Concrete plinths are used as foundations to raise the Wikkelhouse 50cm from the ground. This prevents the structure from getting damaged by rising soil moisture, and it can even withstand floods.

Exterior view.

CARDBOARD SCHOOL

ARCHITECT/INVENTOR: Cottrell & Vermeulen Architecture, Buro Happold
LOCATION: Westcliff-on-Sea, UK
YEAR: 2001
USE: School
CONSTRUCTION TYPOLOGY: Plate structure
AREA: 90m²
PLANNED SERVICE LIFE: approx. 20 years

Schematic exterior view.

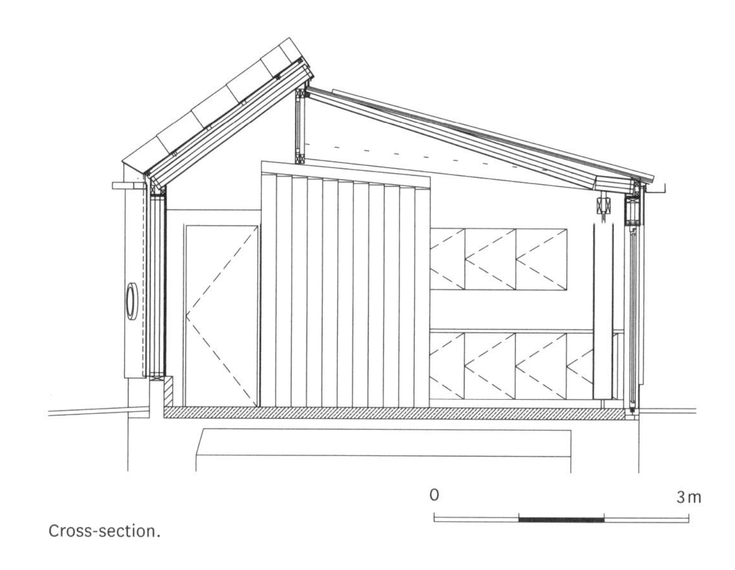

Cross-section.

The Cardboard School was designed to add more common rooms to Westborough Primary School in Westcliff-on-Sea in the UK. Since this extension was conceptualised for a limited period of 20 years, the building authorities approved an experimental paper building designed by architects Cottrell & Vermeulen in collaboration with engineering firm Buro Happold. The planning focused on a resource-saving and degradable construction so that it would not leave too large an ecological footprint when dismantled. For this reason, the designers chose paper as the main construction material – primarily made from waste paper and easy to recycle. The building is considered the first permanent paper building built in Europe. The school community was involved in the planning process and preparations, as the children collected cardboard boxes and helped with the design and development. The aim was to familiarise the children with the topic of sustainable building and to create a sense of personal responsibility. In addition to the paper, most of the other building materials were recycled too. The foundations were designed in such a way that they could be reused in the event of a failure of the paper structure.

It took two years from the initial idea to the completed building. The first year was spent researching and planning, followed by six months for full-size prototype construction and another six months for the construction of the building. In total, 21 different manufacturers and suppliers supported the project by providing free material, labour or expertise. Further financial support was

Exterior view: front.

Interior view: multi-functional space.

Exterior view: rear side with folding doors.

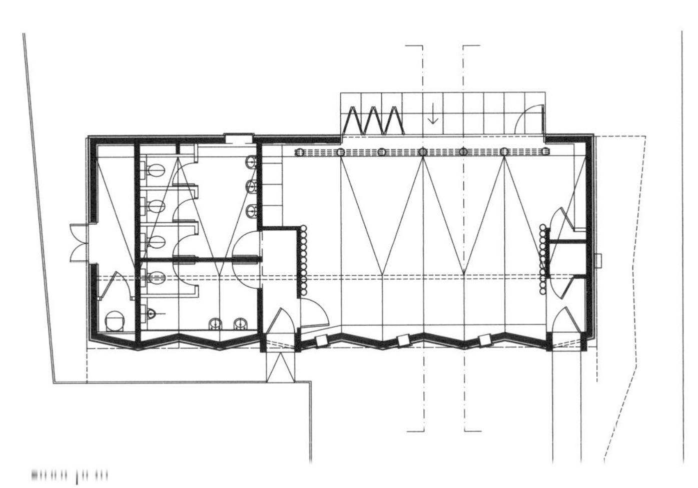

From Box to Building

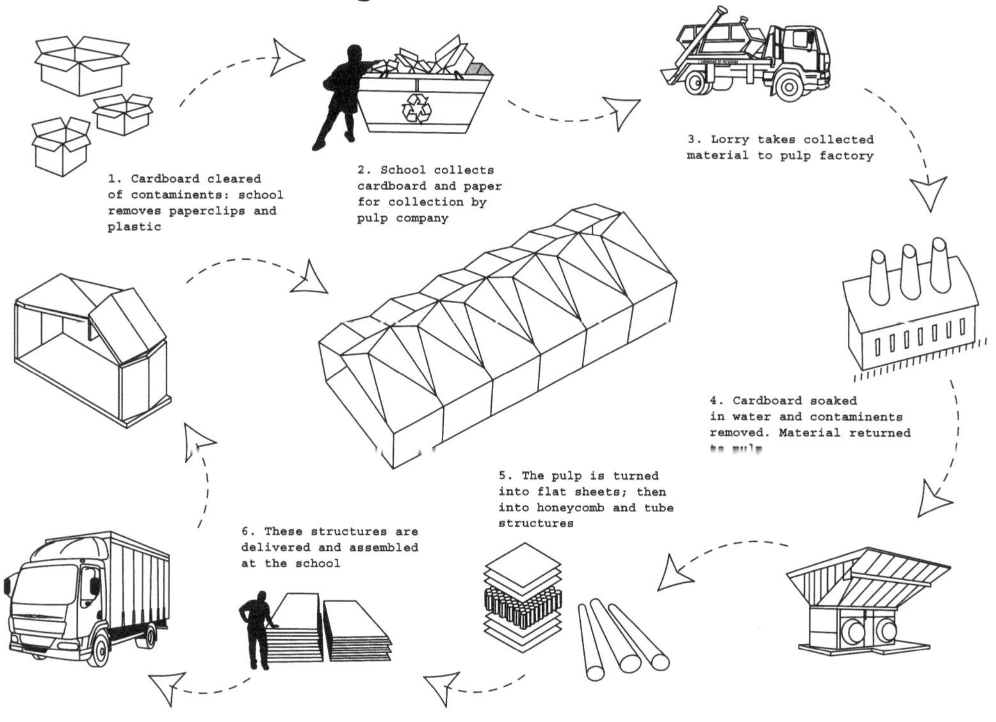

Recycling diagram.

provided by the research partners, the Cory Environmental Trust in Southend-on-Sea, and the DETR through the Partners in Innovation programme. The total cost amounted to £177,157.
At the start of the project, Buro Happold carried out a series of tests on water and fire resistance, strength, creep and durability. One result of these tests was that the compressive strength values were increased by about 10% to avoid deformation due to material creep.
Fire tests showed that a 5mm thick untreated solid board charred rather than burned when exposed to a flame-thrower, forming a natural fireproof barrier. Another finding was that structurally effective paper tubes had to be protected against moisture and considerable temperature fluctuations, resulting in the application of a water- and fire-proof layer. The test results led to a building with paper tubes forming the load-bearing structure and corrugated multi-wall boards forming the enclosing envelope. Two interior walls are composed of 11 juxtaposed paper tubes that support the wooden truss construction of the roof. Another seven paper tubes stand spaced in a row to carry the loads from the roof. The individual layers of the wall and roof panels consist of alternating cardboard panels of different thicknesses: between the outer layers of 4mm thick solid board are a total of three 50mm thick honeycomb boards, each in turn separated by a layer of solid board. This structure achieves a U-value of 0.3 W/(m²·K).
The panels are glued together and accurately fitted to lie in a wooden frame. For manufacturing reasons, their size is limited to a maximum height of 2.7m and a width of 1.5m. To minimise the risk of contact with moisture and water, the panels were equipped with a vapour-proof coating on the inside and waterproof but diffusion-open construction paper on the outside. This allows any room humidity penetrating the paper material to escape to the outside.
On the inside, the layer of 4mm thick solid board adequately protects the panels. On the outside, however, a cladding of 16mm thick fibre cement panels was considered necessary to protect the wall and roof panels from point loads such as hailstorms and to prevent damage from playing pupils or similar.
Four months after the building was erected, deformations could be seen here and there on the cardboard tubes.

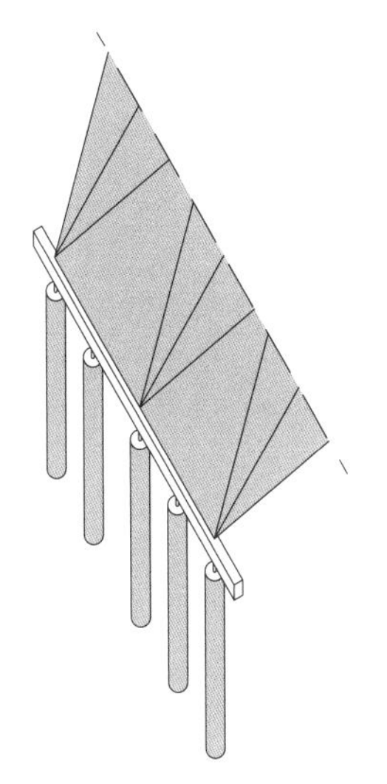

Load-bearing structure of the Cardboard School: cardboard tubes support the wooden truss structure of the roof.

Detail of the connecting elements.

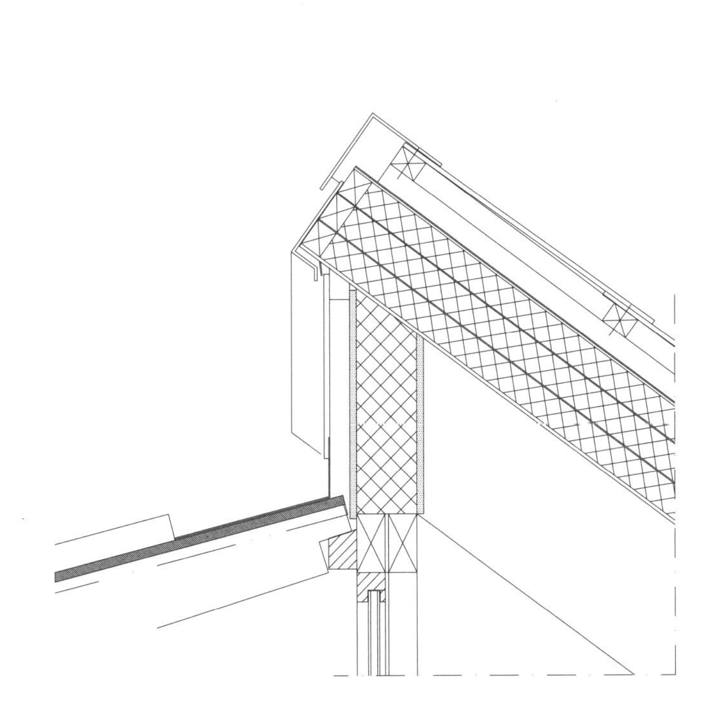

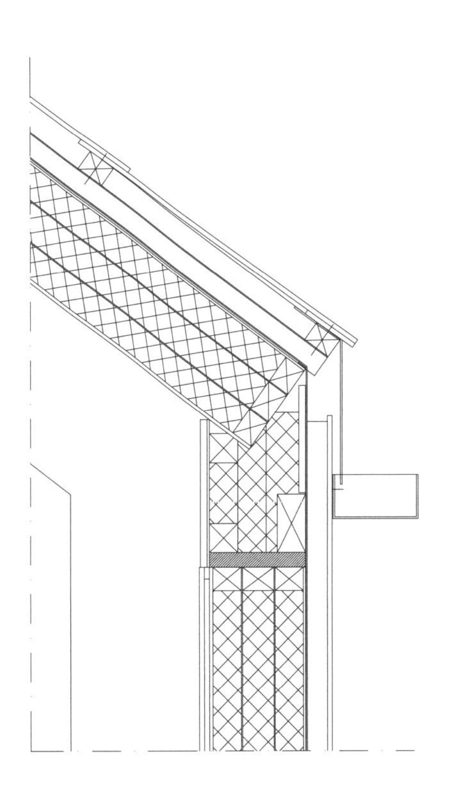

Section detail ridge and eaves.

It turned out that the lateral movement at the top of the wall was caused by the paper tubes drying out, which changed their dimensions. In order to reinforce the outer walls, several partition walls were installed on the inside – after which no further movements could be detected.

In addition, a deflection of approx. 10–15mm was observed in the paper tubes due to subsequent drying. The original goal of constructing a building that could be 90% recycled was thwarted by the concrete foundation and the necessary wooden elements.

Nevertheless, the building has received a number of awards for its innovative and sustainable use of materials, including the RIBA Award 2002, the RIBA Stephen Lawrence Prize 2002, the RIBA Journal Sustainability Award 2002 and a Civic Trust Awards Commendation (2002).

CARDBOARD HOUSE

ARCHITECT/INVENTOR: Peter Stutchbury and Richard Smith, Ian Buchan Fell Housing Research Unit of the University of Sydney, Stutchbury and Pape Architects
LOCATION: Sydney, Australia
YEAR: 2004
USE: Living
CONSTRUCTION TYPOLOGY: Plate structure
AREA: 32.4m² and 7.2m² mezzanine
PLANNED SERVICE LIFE: Temporary

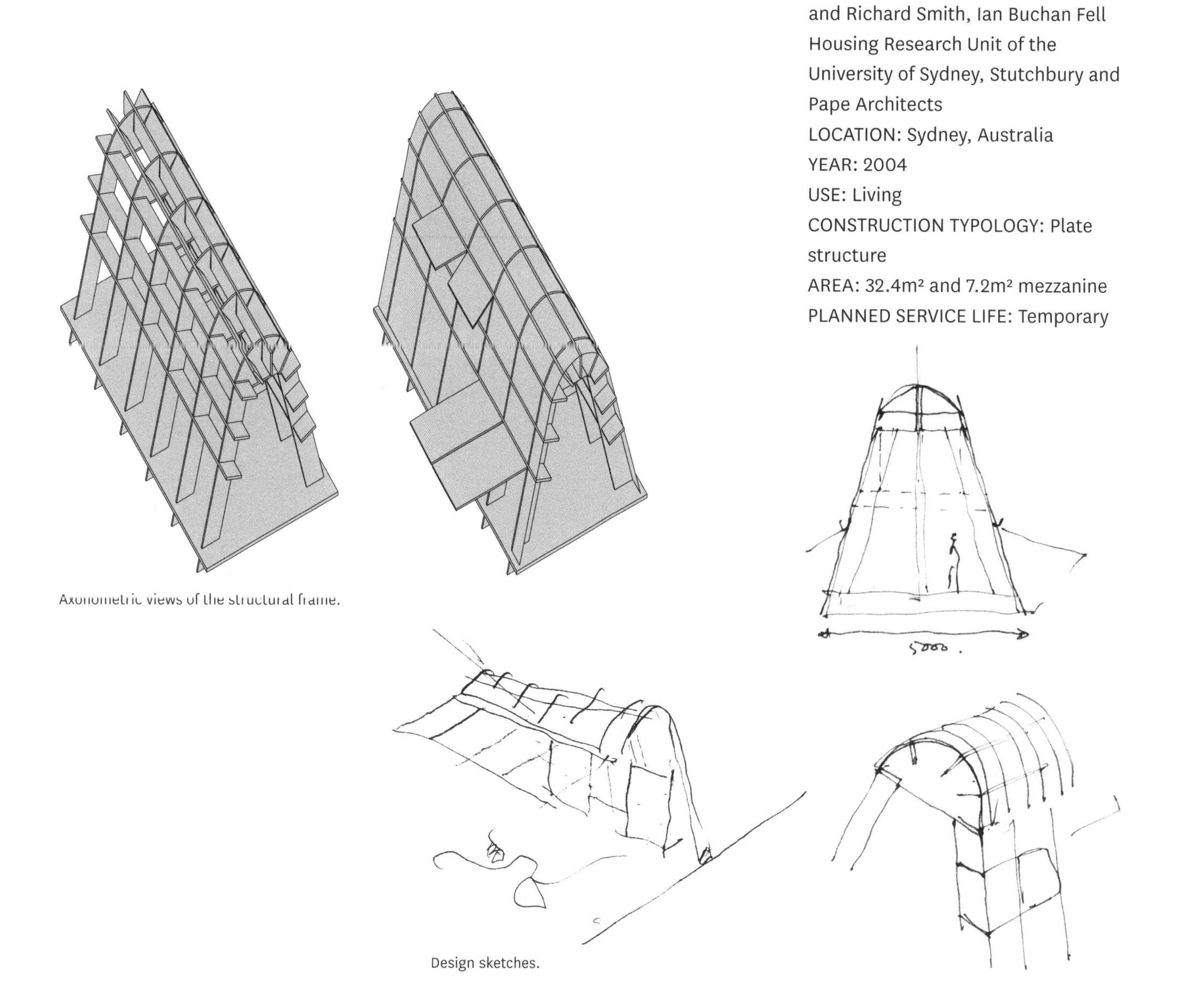

Axonometric views of the structural frame.

Design sketches.

The Commonwealth Government in Australia declared 2004 the Year of the Built Environment (YBE). One of the main events of the YBE was the House of the Future exhibition, organised by the Australian State Government of New South Wales. The exhibition presented six concept houses made of very different materials: there was one house each made of concrete, cardboard, glass, clay, steel and wood. The Cardboard House was designed by Peter Stutchbury and Richard Smith of Stutchbury and Pape Architects in collaboration with the Ian Buchan Fell Housing Research Unit at the University of Sydney. The idea was to create a cost-effective, lightweight, easy to transport and assemble, and recyclable construction. The temporary house was to consist of 85% recycled materials and be fully recyclable after use.
The load-bearing structure of the house was formed by six A-shaped arches made of recycled solid board. The arches stood next to each other at a distance of 1.8m and were interlocked with horizontal cardboard spacers. Each arch was composed of two 5.1m × 600mm × 60mm struts and a semi-circular crown.
The horizontal purlins or spacers measured 10.2m × 600mm × 60m. 10mm solid boards were used for bracing, combined with 50mm PET pipes and M12 threaded nylon rods.
At the base, the volume was attached to concrete girders with steel clamps. The

Exterior view.

Interior view.

A covering of high-density polythene textile served as weather protection for the Cardboard House, both for resistance to rain and from moisture at night.

cardboard construction was protected against the weather by a high-density polythene textile that enveloped the structure and was also recyclable. Grey water draining from the cover was collected in underground tanks.
The internal structure of the house consisted of an open, furnished room on the ground floor and a bedroom on the mezzanine.
Service units such as a kitchen and bathroom with a composting toilet completed the facilities. The swing-door panels opposite the service units extended the living space with additional outdoor areas. A photovoltaic system generated 12V solar power.
Since the entire Cardboard House only weighed around 2000kg and came in the form of a flat package when dismantled, a light commercial vehicle was sufficient for transport. Assembly required a small scaffolding but no other special equipment. It took two people only six hours to assemble the house. It stood for a total of one and a half years in three different locations in Australia.

CLUBHOUSE: RING PASS HOCKEY AND TENNIS CLUB

ARCHITECT/INVENTOR: Nils-Jan Eekhout, Octatube
LOCATION: Delft, Netherlands
YEAR: 2010
USE: Clubhouse with community room and childcare
CONSTRUCTION TYPOLOGY: Space frame made of cardboard tubes
AREA: 128m²
PLANNED SERVICE LIFE: Permanent

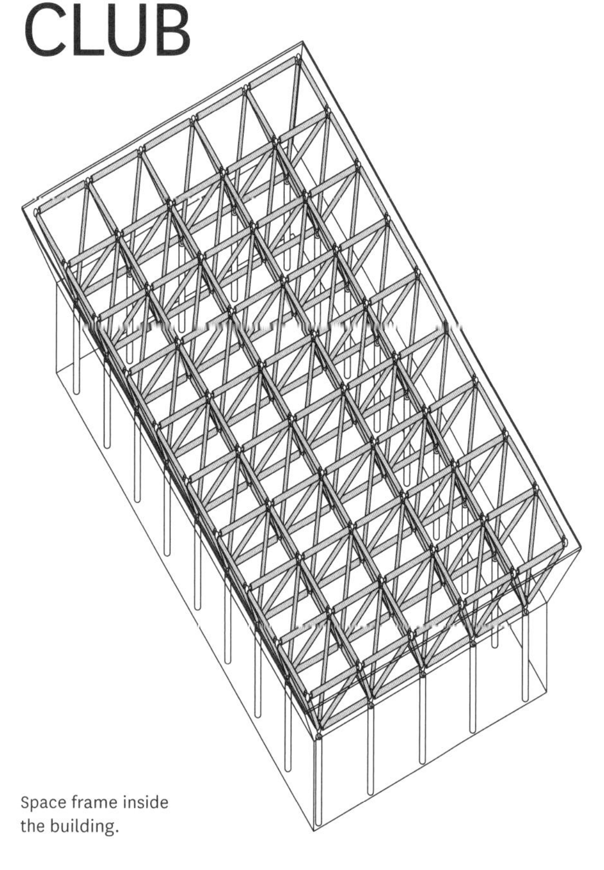

Space frame inside the building.

Tuball node in cardboard tube space frame.

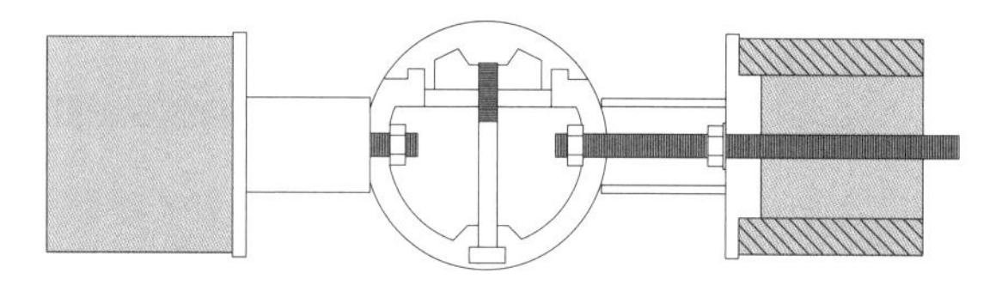

Tuball connection for cardboard tube space frame.

Dutch design and engineering firm Octatube had already collaborated with Shigeru Ban on several projects in which cardboard tubes were used as a structural element. Based on this experience and the "Cardboard in Architecture" research project Mick Eekhout carried out at TU Delft, Nils-Jan Eekhout, technical director of the company, designed a space frame construction for the extension to the clubhouse of the Ring Pass Hockey and Tennis Club in Delft. The construction was pre-assembled in two parts of 8 x 8m each on the ground. The components were then lifted and positioned on the steel columns along the perimeter of the building. Tinted glass panels enclose the space frame and walls, while the roof is covered with corrugated sandwich metal sheets. The individual cardboard tubes of the space frame are connected to each other with spherical Tuball nodes. For this purpose, flange plates with openings for the tension rods were attached at each end. The rods are threaded on either end and run through the cardboard tubes. They are fastened to the two flange plates with nuts, which means that no additional bolting was necessary.

Space frame inside the building.

Exterior view.

With this solution developed by Octacube, the tubes are only exposed to compressive forces. The rubber-sealed flanges prevent moisture from entering the cardboard tubes. The roof construction meets all fire protection regulations because of the cardboard material's high density and a carbon layer that forms on the surface of the cardboard tubes in the case of a fire. Since the cardboard framework itself is not exposed to the weather, only the relative humidity in the room can influence or endanger the construction from a building physics point of view. To identify a means to counter this risk, the surfaces of the cardboard tubes were treated in three different ways: some were covered with heat-shrinkable polythene sleeves, others were painted on the inside and outside, and a few were simply left untreated. The construction is regularly checked for its structural integrity.

PH-Z2

ARCHITECT/INVENTOR: Dratz & Dratz Architects
LOCATION: Essen, Germany
YEAR: 2010
USE: Multi-functional
CONSTRUCTION TYPOLOGY: Solid construction from waste paper masonry
AREA: 185m²
PLANNED SERVICE LIFE: Temporary (three years)

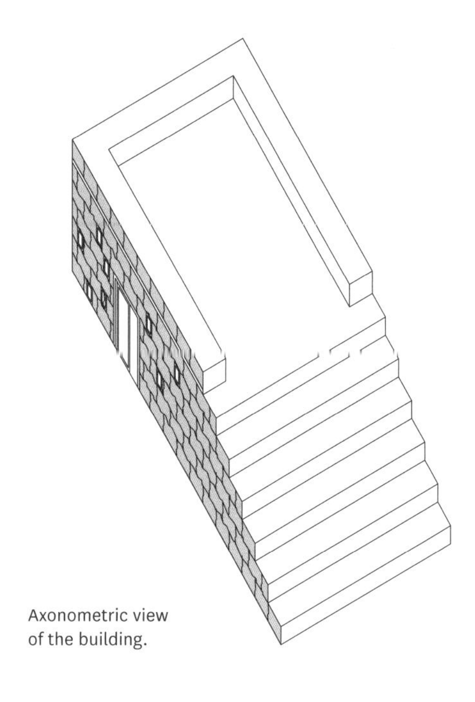

Axonometric view of the building.

Exterior views.

The name PH-Z2 stands for "Papierhaus Zukunftsweisend 2", i.e. Paper House Future Oriented 2. It was developed from a project that was awarded the first prize in a competition on "mobile working spaces". This single-storey building of 30m length comprised a multi-functional room and secondary rooms to serve as a cinema, concert hall, and club and lecture areas with just under 200 seats. The multi-functional room was about 30 × 6.6m in plan and 4.8m high; it was equipped with a bar, storage, toilets and secondary rooms.

Architects Dratz & Dratz developed and patented building blocks made of compacted waste paper. One compacted building block is approx. 1.4 × 1.1 × 0.8m in size and weighs half a tonne. Some 550 blocks were needed for the building, resulting in a total weight of 275 tonnes.

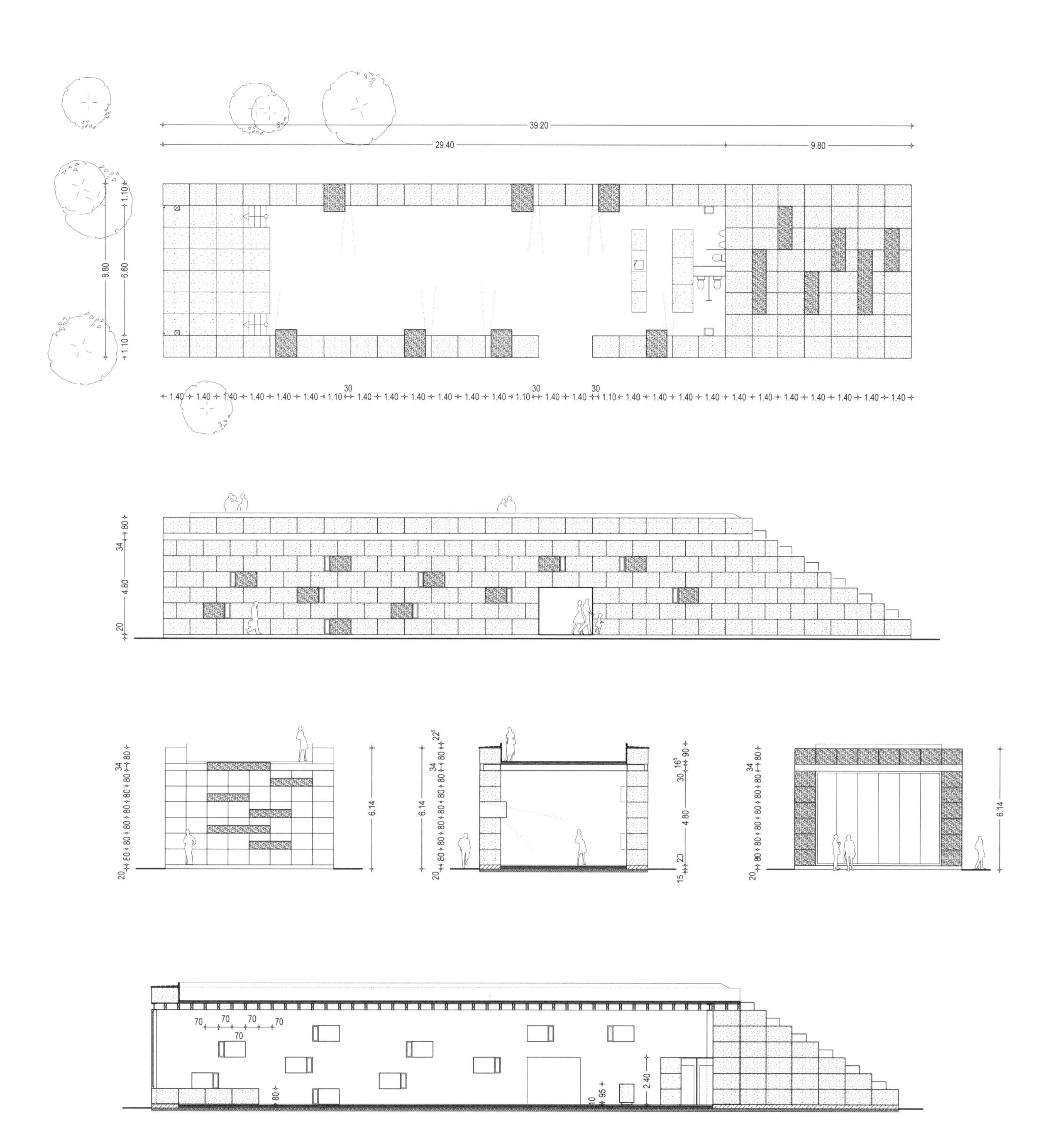

Floor plan, side elevation, view of stepped wall, cross-section, view of forest window, longitudinal section.

Interior views.

Compressing the paper blocks means the material is highly compacted and therefore features higher strength values than normal as well as improved fire-retardant properties.
In addition, the size of the blocks creates an insulating effect. The blocks are impregnated on the outside for moisture protection.
The planning aimed to create a circular building construction sequence, in which the material can be recycled after use.

The essential features of the construction – load-bearing, insulating, inexpensive and recyclable – remain clearly visible on the outside. Accordingly, this construction method was patented.
The paper blocks – simple, cheap and quick to produce – were used for the first time in a building in Essen. The construction costs amounted to 170,000 euros.
For the purpose of strength testing, IKT (Institute for Underground Infrastructure gGmbH) carried out load tests.

The paper blocks themselves were tied with steel straps. According to the European list of standard grades of paper and board for recycling, they must consist of 70% solid board and packaging paper to be considered recyclable.
Fire safety was tested by means of experiments. Due to the density of the paper blocks, they typically smouldered but did not catch fire. Despite these tests, the hall was destroyed in a fire.

Waste paper baling machine.

Load test in a hydraulic press.

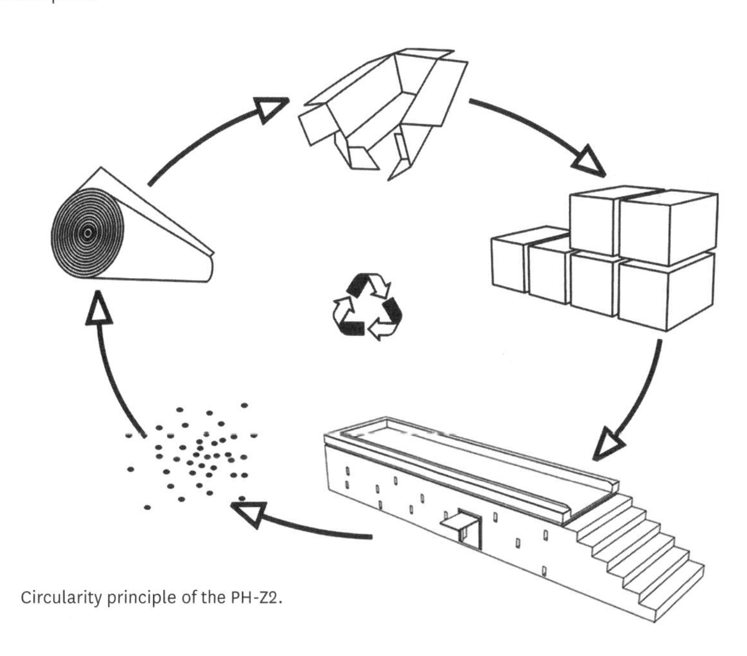

Circularity principle of the PH-Z2.

Stacked paper bales during assembly.

Cross-sectional examination to determine water penetration.

INSTANT HOME

ARCHITECT/INVENTOR: Technical University of Darmstadt, Department of Architecture, Mechanical Engineering and Chemistry; Department of Plastic Design
LOCATION: Darmstadt, Germany
YEAR: 2012
USE: Research project
CONSTRUCTION TYPOLOGY: Folded structure made of sheet materials
AREA: 20m²
PLANNED SERVICE LIFE: Temporary

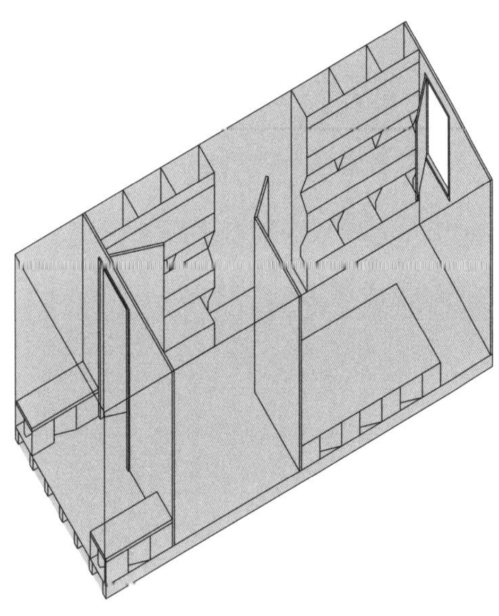

Axonometric view.

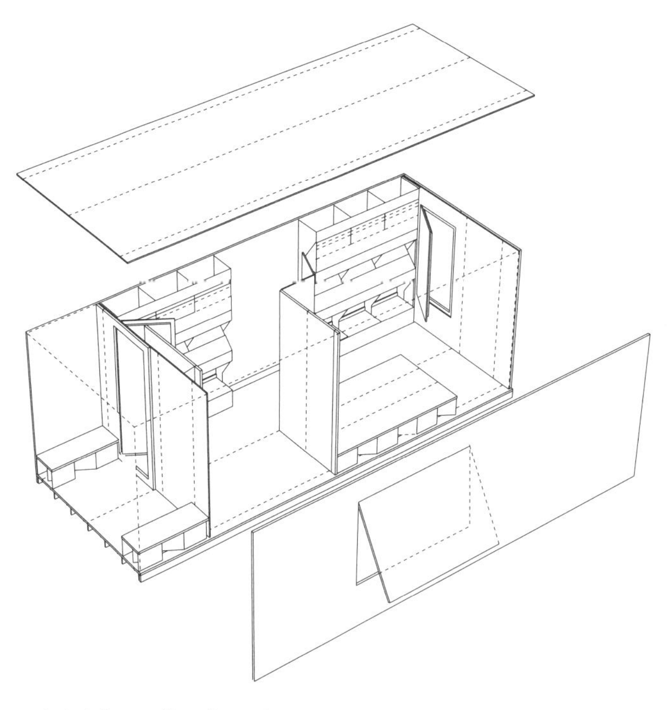

Exploded view cardboard container.

"Instant Home" is the name of the interdisciplinary research project of the Departments of Chemistry, Architecture and Mechanical Engineering at the TU Darmstadt. The project is the first interdisciplinary project in the field of building with paper and is considered a cornerstone for other research projects, including "BAMP! –Building with paper".
The cooperation of the different disciplines was intended to lead to a cross-disciplinary research project.
The idea was already sketched out in the early project development phase; in the course of the work, it was then concretised to the effect that only completely biocompatible paper materials were to be used.
The project started in 2011/2012, aimed at developing rapidly realisable shelters for global natural disasters. The starting point was an interdisciplinary student competition with the task of developing a concept for a simple dwelling made of paper materials.
Many designs laid the focus on foldable structures to achieve a simple and quick assembly and to exploit the main advantage of paper – its lightness.
The Cardboard Container House concept

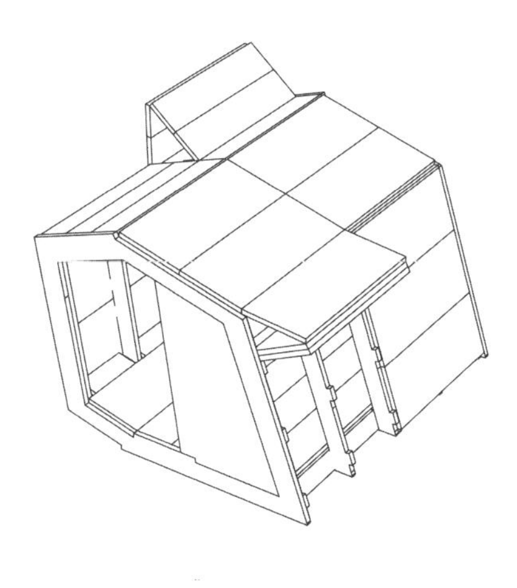

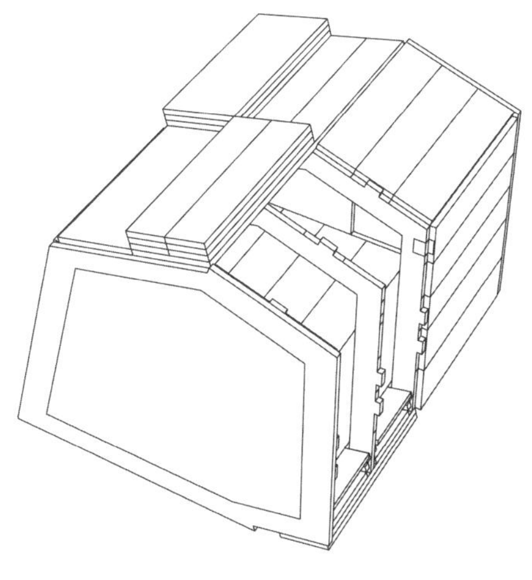

Schematic drawing.

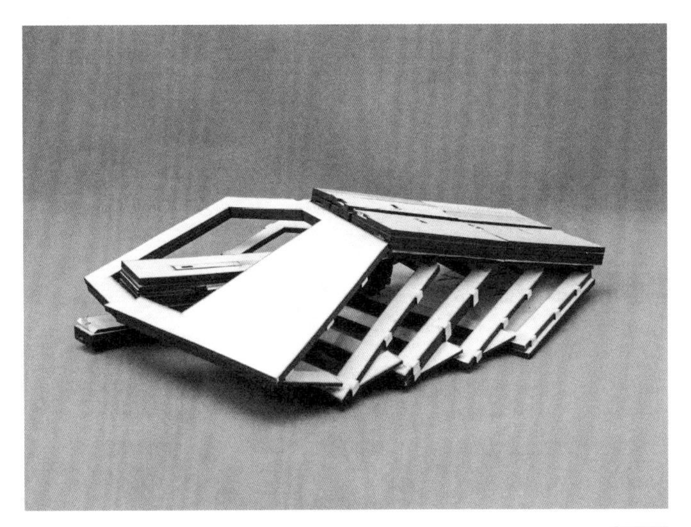

Models of the Instant Home.

was selected for further development, and a prototype was planned to test the feasibility of series production and thus determine large-scale deployment in disaster areas.
When folded, the simple basic shape results in a space-saving cargo item whose size is tailored to the internal dimensions of an ISO container. Panels or ribs and a centrally placed

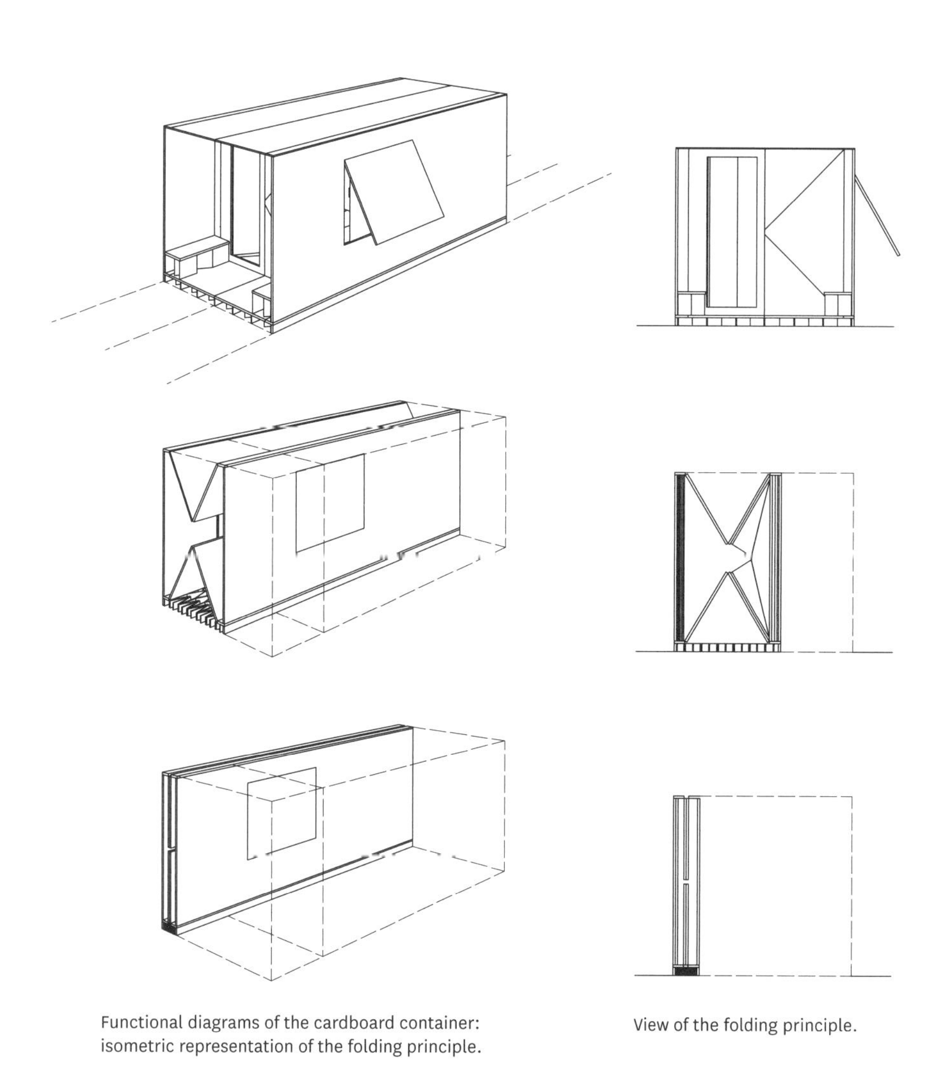

Functional diagrams of the cardboard container: isometric representation of the folding principle.

View of the folding principle.

Partial demonstrator.

room divider brace the envelope and stabilise the walls and ceiling. Vertical ribs on one side of the building are connected to each other with horizontal ribs. They improve the stability of the longer outer walls and prevent deformation due to buckling.
This structural necessity gave rise to an unexpected design element: a wall shelf. The one room concept was assessed in the longitudinal direction of the 2.28 × 5.80m plan and 2.20m high module. The exterior components of the house were built symmetrically in the longitudinal direction. When packed, the dimensions are reduced to 0.31 × 5.80 × 2.20m. This means that a total of six modules can be transported with a standard ISO container.
A special feature of the building is its folded joints. The structure fulfils the basic principle of the Instant Home, according to which the house should be easy to assemble and dismantle and easy to transport. A final demonstrator of the Container House was built as a partial model in 2014.

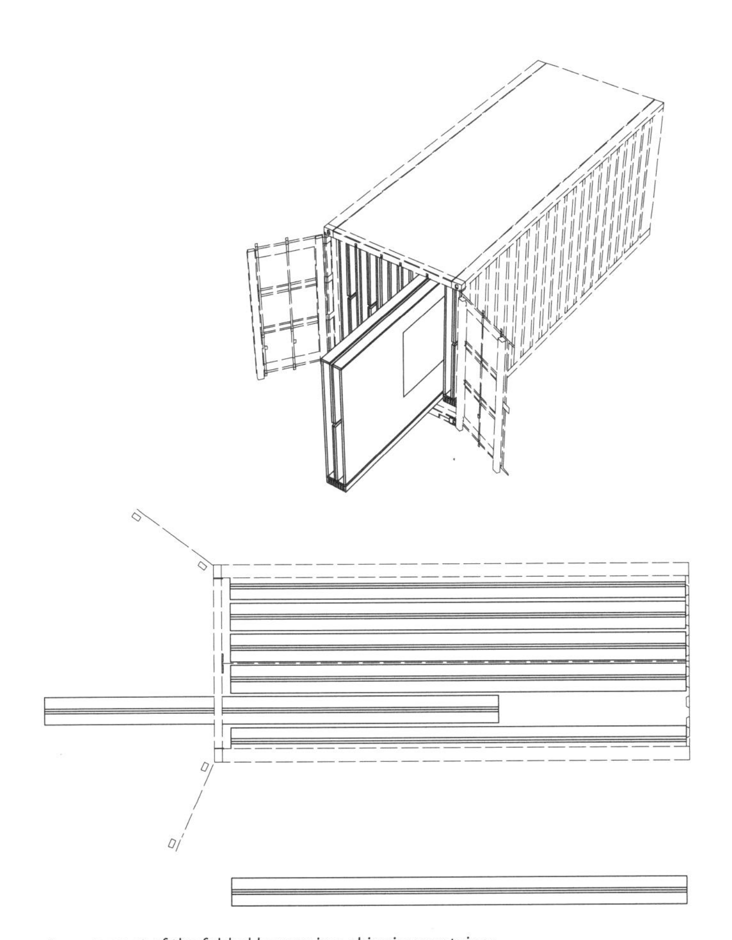

Arrangement of the folded houses in a shipping container.

Illustration of the interior and the floor construction.

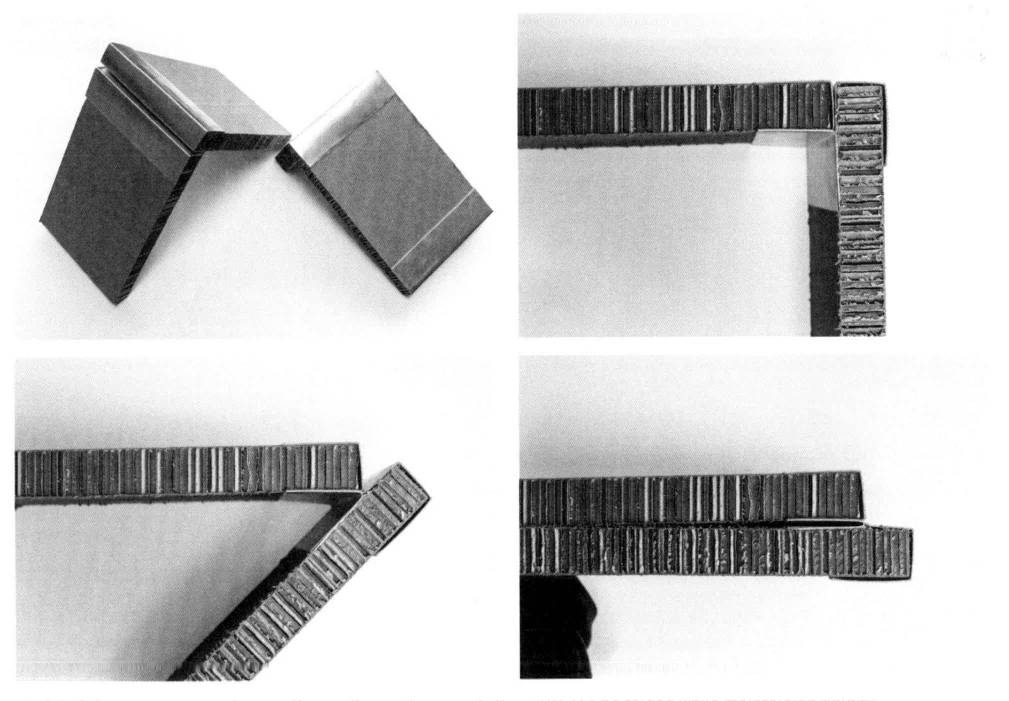

Model of the corner connection: with an adhesive tape joint, the walls can be folded while maintaining rigidity.

STUDIO SHIGERU BAN KUAD

ARCHITECT/INVENTOR: Shigeru Ban
LOCATION: KUAD Campus, Kyoto, Japan
YEAR: 2013
USE: University studio
CONSTRUCTION TYPOLOGY: Shell (arch) construction made of cardboard tubes
AREA: 142m²
PLANNED SERVICE LIFE: Permanent

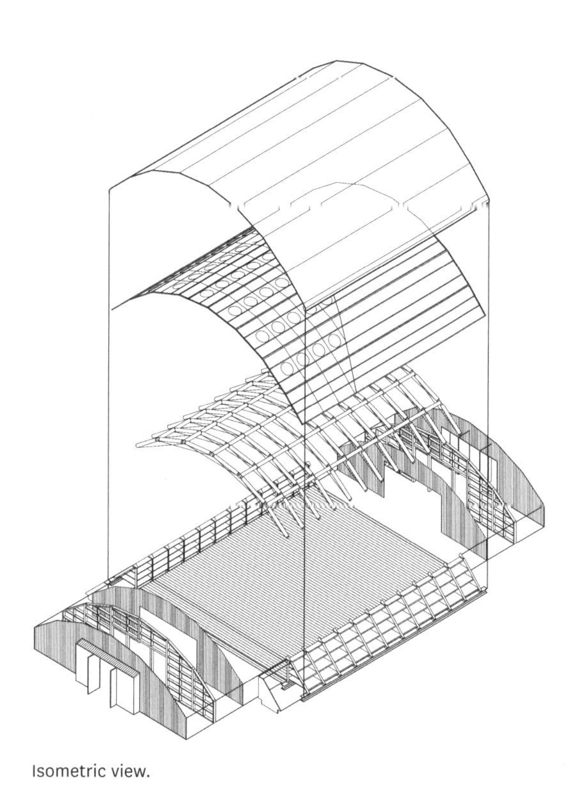

Isometric view.

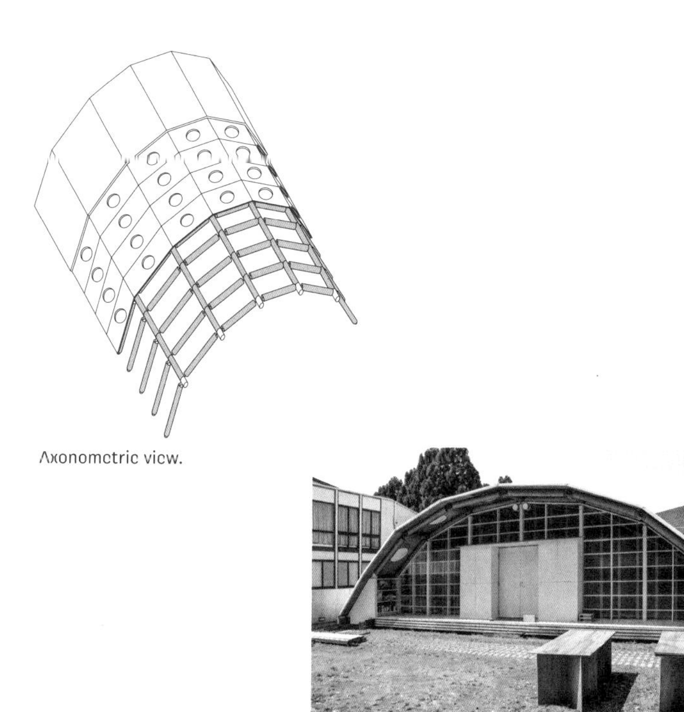

Axonometric view.

Exterior view.

In 2011, Shigeru Ban was appointed professor at Kyoto University of Art and Design (KUAD). Two years later, he decided to build a temporary studio on the KUAD campus together with his students – a space for individual and team work, study, design and model making. The studio consisted of an arch-shaped shell construction made of cardboard tubes. It was based on an idea that Ban had already used in other projects such as the Paper Arch Dome in 1998, the Ban Lab at Keio University in 2003 or the temporary office of Shigeru Ban Architects on the sixth-floor terrace of the Centre Georges Pompidou in Paris in 2004.

The studio on the KUAD campus has a floor area of 11.7 × 12.1m and is founded on a concrete slab.

Along each of the two sides of the shell runs a 250 × 250mm steel H-beam on a concrete base, to which the cardboard tube arches are attached. The 12 arch supports are each made of 1.86m long cardboard tubes with an inner diameter of 170mm and a wall thickness of 3.5mm. Five rows of transverse cardboard tubes (length per tube = 850mm) of the same type connect the arches, creating a shell grid of 2.2 × 1.2m. This cardboard tube grid covers a substructure of wooden slats to which plywood panels are attached, providing the structural rigidity of the shell. The circular holes (ø 750mm) cut into the panels allow daylight to

Steel connection detail: at the end of each tube are wooden inserts with steel plates that connect the elements by means of two pre-stressed steel rods each.

Arch construction made of cardboard tubes, here with metal nodes.

Model of the steel connection.

shimmer through the translucent insulation layer and reduce the weight of the wooden panels.
In addition, the students were able to reach and fasten the next elements for the assembly without having to climb onto the roof. The weather-protective roof cladding is a PVC sheet laid over the polystyrene insulation boards, which in turn rest on the plywood boards. The students use the cross-sectionally triangular supports at the edges of the semi-circular roof as storage space. The gable walls are composed of wooden frames clad with translucent corrugated plastic sheets to allow daylight to enter. The entrance in the middle of each gable wall is formed by two raised storage boxes with doors between them.
In contrast to earlier projects with arch constructions, in which the cardboard tube joints were made of wood, Shigeru Ban used metal nodes for his studio project. The cardboard tubes making up the arches were sealed with wooden inserts and covered with steel plates. Inside each tube are two pre-stressed steel rods, each forming a metal joint at the ends to stiffen the structure and prevent deformation in the event of sudden load changes due to wind or snow. The transverse cardboard tubes are connected to each other with wooden pegs.

HOUSE OF CARDS

ARCHITECT/INVENTOR: Jerzy Łątka (archi-tektura.eu)
CONSTRUCTION: Julia Schönwälder
TECHNICAL CONSULTATION: Marcel Bilow
LOCATION: Wrocław, Poland
YEAR: 2016
USE: Temporary housing
CONSTRUCTION TYPOLOGY: Frame construction
AREA: 16m²
PLANNED SERVICE LIFE: Temporary (18 months)

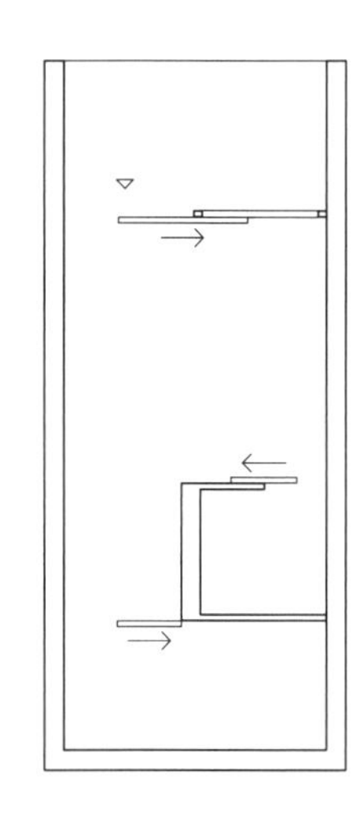

Axonometric view and floor plan of the paper structure.

House of Cards belongs to the category of emergency shelters that provide refuge for people in difficult living situations. However, the temporary cardboard frame construction is also suitable as a cost-effective and environmentally friendly housing solution anywhere in the world.
The project was developed based on previous research on paper-based housing units ("TECH" = Transportable Emergency Cardboard House) and elaborated in close cooperation with structural consultant Julia Schönwälder. House of Cards won the FutuWRO competition as part of the European Capital of Culture Wrocław 2016 programme.
The concept comprised two residential units with different spans of 2.6m and 4.3m, respectively. The structural design of the two units consisted of a T-shaped cardboard frame construction, filled with repeating prefabricated sandwich panels made of honeycomb cardboard (1.1 × 2.2m) and roof panels.
The latter varied depending on the roof pitch and size. The longitudinal dimensions of the units can be adjusted, depending on the number of components.
The units can be arranged differently and thus suit different types of residents, such as a group of workers or several families.
The smaller unit was exhibited on Wrocław's historic market square during the European Capital of Culture 2016 programme. The project was implemented in cooperation with the Wrocław University of Science and Technology, whose students were involved in the

House of Cards on the market square in Wrocław, 2016, exterior view.

construction. Marcel Bilow from the Faculty of Architecture at TU Delft acted as technical advisor during the production process.

The prefabricated 2.6 × 6m wooden floor of the House of Cards rested on a 14 × 14cm foundation beam with wooden feet. Walls and roof frames consisted of 200 × 100mm solid cardboard T-profiles with a wall thickness of 10mm. Honeycomb composite panels were installed between the frame elements to stabilise the overall structure. The panels were installed from the outside so that they could be maintained or replaced at any time. The structural volume consisted of 70% cardboard elements.

A polythene film protected the outside of the honeycomb cardboard wall elements, supplemented by a self-adhesive PVC film.

The round windows were made with large cardboard tubes sealed with liquid rubber. On the roof, photovoltaic modules in combination with a battery provided enough solar power to supply an LED lighting system with energy for 48 hours in bad weather.

House of Cards was presented on the market square in Wrocław for a fortnight and then transported to the campus of the Faculty of Architecture at the Wrocław University of Science and Technology (WUST), where the construction was monitored for the next 18 months.

After this time, the building had to be demolished due to the redevelopment of the campus.

Interior view.

Detail of the connection between frame element and honeycomb wall panel.

TECH 04

ARCHITECT/INVENTOR: Jerzy Łątka (archi-tektura.eu)
LOCATION: Wrocław, Poland
YEAR: 2018
USE: Emergency shelter and other housing
CONSTRUCTION TYPOLOGY: Plate structure
AREA: 16m²
PLANNED SERVICE LIFE: Temporary

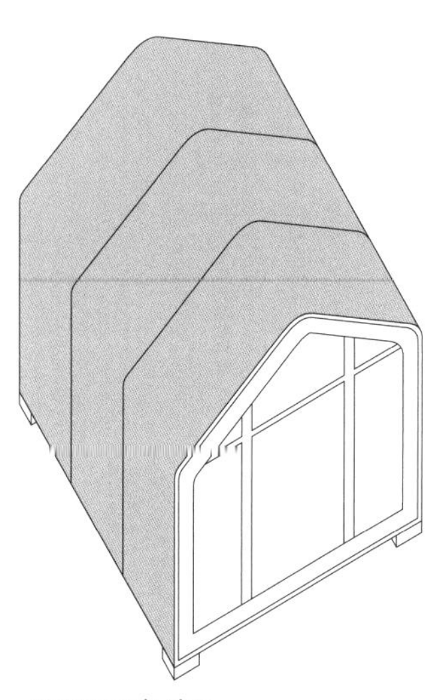

Axonometric view.

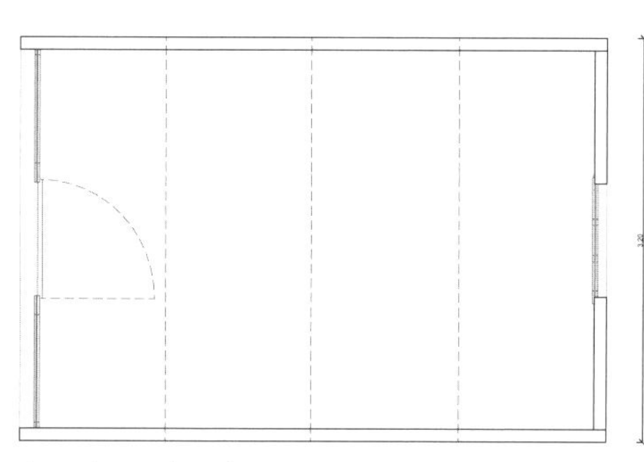

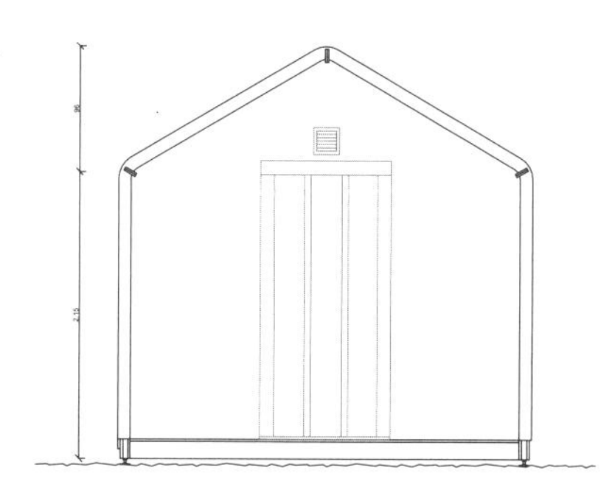

Floor plan and section.

The fourth generation of TECH was developed by Agata Jasiołek and Prof. Mick Eekhout. The project was built by students from Wrocław University of Science and Technology. The flat, prefabricated elements of the 4.8 × 3.2m house were transported to the construction site and assembled on site using basic tools such as a drill, screwdriver and hammer.

The construction is a composition of prefabricated wooden floor elements, long foldable wall-roof-wall components and gable walls. The wall-roof-wall sandwich panels, each measuring 7.1 × 1.25m, consist of a honeycomb panel core to which four layers of corrugated board are applied on both sides.

On the outside, these foldable panels are sealed with aluminium sheeting and, on the inside, with self-adhesive foil. According to the calculations, the building components have a thermal conductivity coefficient (U-value) of 0.52 W/(m² · K) for the walls and 0.54 W/(m² · K) for the roof. Two notches allow for the targeted folding of the long wall-roof-wall panel. During assembly, one edge is first fixed to the wooden floor; then, it is folded and fixed to the other side of the wooden floor. A second component is mounted next to the first. The aluminium sheets overlap and act as connecting strips. Finally, the front and rear gable walls are attached and fixed with screws.

The finished house was monitored for one and a half years for its structural integrity and thermal insulation performance. Investigations with a thermal camera showed the thermal impermeability of the connections between the TECH 04 components.

Exterior view: entrance.

aluminium 0.6mm
corrugated cardboard 4 × 7mm
honeycomb panels 2 × 25mm
corrugated cardboard 4 × 7mm
self-adhesive foil

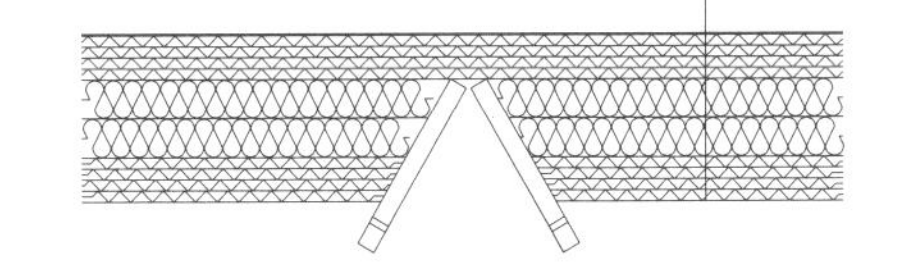

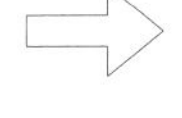

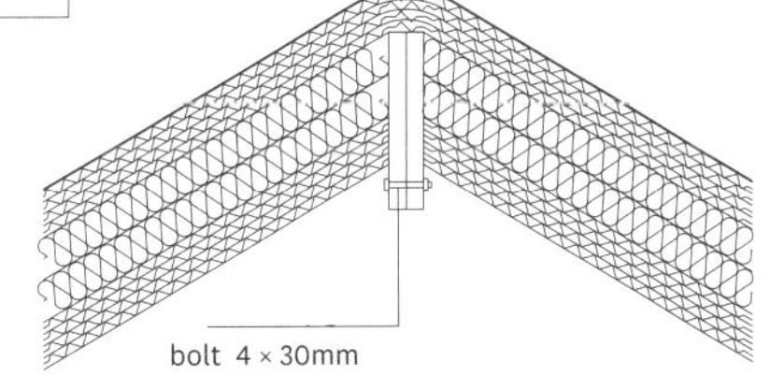

Detail of the building envelope and folding process.

Exterior view: rear.

HOUSE 01

ARCHITECT/INVENTOR: BAMP!, Technical University of Darmstadt
LOCATION: Darmstadt, Germany
YEAR: 2018
USE: Research demonstrator
CONSTRUCTION TYPOLOGY: Skeleton construction
AREA: approx. $16m^2$
PLANNED SERVICE LIFE: Temporary

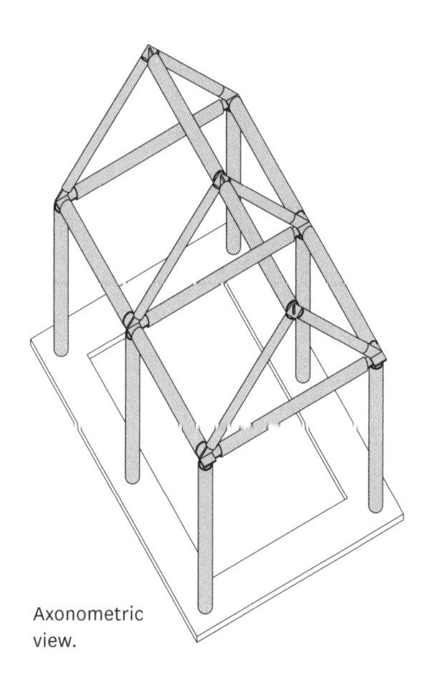

Axonometric view.

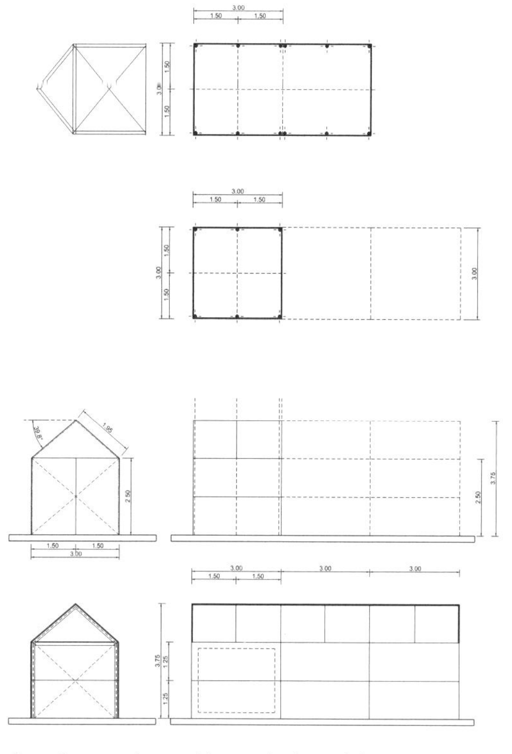

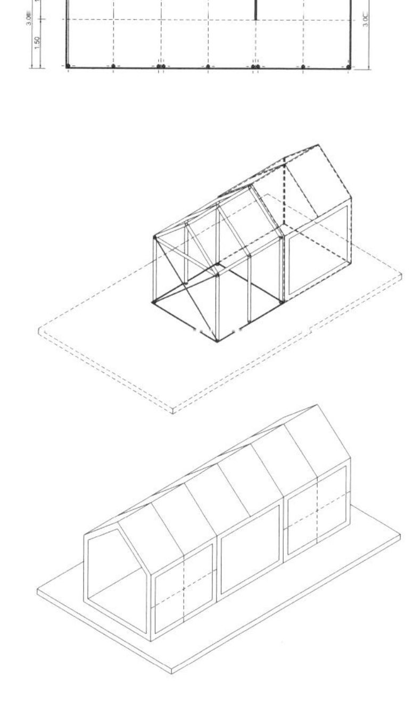

Floor plans, sections and isometric views of the support structure.

House 01 was created in 2018 as the first large-scale demonstrator of the research project “BAMP! – Building with paper”. The interdisciplinary team of scientists began their research work by planning architectural constructions and details for demonstrators made of paper materials. It was hoped that these experimental set-ups would provide decisive insights into the usability of paper and cardboard materials currently available in the construction industry.
In the sense of a systematic approach to paper-made structures, the focus was initially on considerations of details and superstructures with paper-based production and design methods. Construction principles and details were tried out on a 1:1 scale, and a theoretical-conceptual series of designs for house demonstrators was defined based on different construction or building methods. In addition, different requirements were placed on planning and construction.
Two initial houses were developed based on commercially available paper materials: House 01 and House 02 **(» pp. 122–123)** as simple house archetypes, with a rectangular floor plan and a gable roof with 35° to 45° pitched roof surfaces.
A skeleton construction method was chosen for House 01. The load-bearing

Demonstrator under construction: the tubes of the skeleton construction are secured with plug-in connections. In this case, multi-layered honeycomb panels form the outer skin.

system of the demonstrator building consists of paper tubes; the building envelope of honeycomb panels with connecting elements made of MDF, which also anchor the house in the foundation made of wood-based materials.
The theoretical and practical development of House 01 focused on the dimensioning with regard to dead loads and aimed at simple elementarisation of the construction and thus at a rapid build with uniform materials.
During the planning of House 01, special attention was paid to the development of the nodes in the building structure as well as to the dimensioning of the columns and beams. In terms of joining technique, the BAMP! team opted for plug-in connections with light cross-pins that connect the sleeves by means of friction. This meant that there was no need for conventional connectors such as screws, nails or adhesives.
The selected skeleton construction required that the tubes are connected at the ends – i.e. in the resulting nodes – in three to four directions by means of plug-in connections, thus closing and stabilising the system.
The construction was conducted iteratively: first, the supports had to be clamped into the floor; then, the other pre-assembled frames were successfully added and fixed in place using a locking plug-in connection. The overall assembly of the tube columns and beams forms the main structure, encased in an outer layer of multi-layered honeycomb panels attached to the tubes.
For the concept definition, actual test data of the tubes was fed into the simulation programme to calculate the structural integrity of the building. The focus lay on gaining fundamental knowledge about the material for calculating and dimensioning other paper-based elements.
The development and calculation of the design brought about important insights into the tube structures, especially with regard to windings and bonding, as well as their relevance in terms of structural properties.
Based on House 01, the team developed nodes that allowed for the structure to be joined and assembled in the relatively small space of the research studio, meaning under cramped conditions. The "key-lock" plug-in system is a model example. In the follow-up to the realisation and evaluation of the project, the BAMP! team expanded the research on alternative nodes » **chapter 4, section on hybrid connections, pp. 61–63.**

HOUSE 02

ARCHITECT/INVENTOR: BAMP!, Technical University of Darmstadt
LOCATION: Darmstadt, Germany
YEAR: 2019
USE: Research demonstrator
CONSTRUCTION TYPOLOGY: Solid construction
AREA: approx. 16m²
PLANNED SERVICE LIFE: Temporary

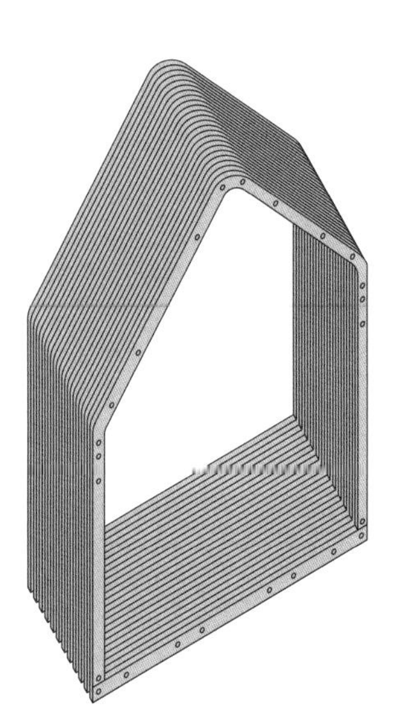

Axonometric view of the outer shell.

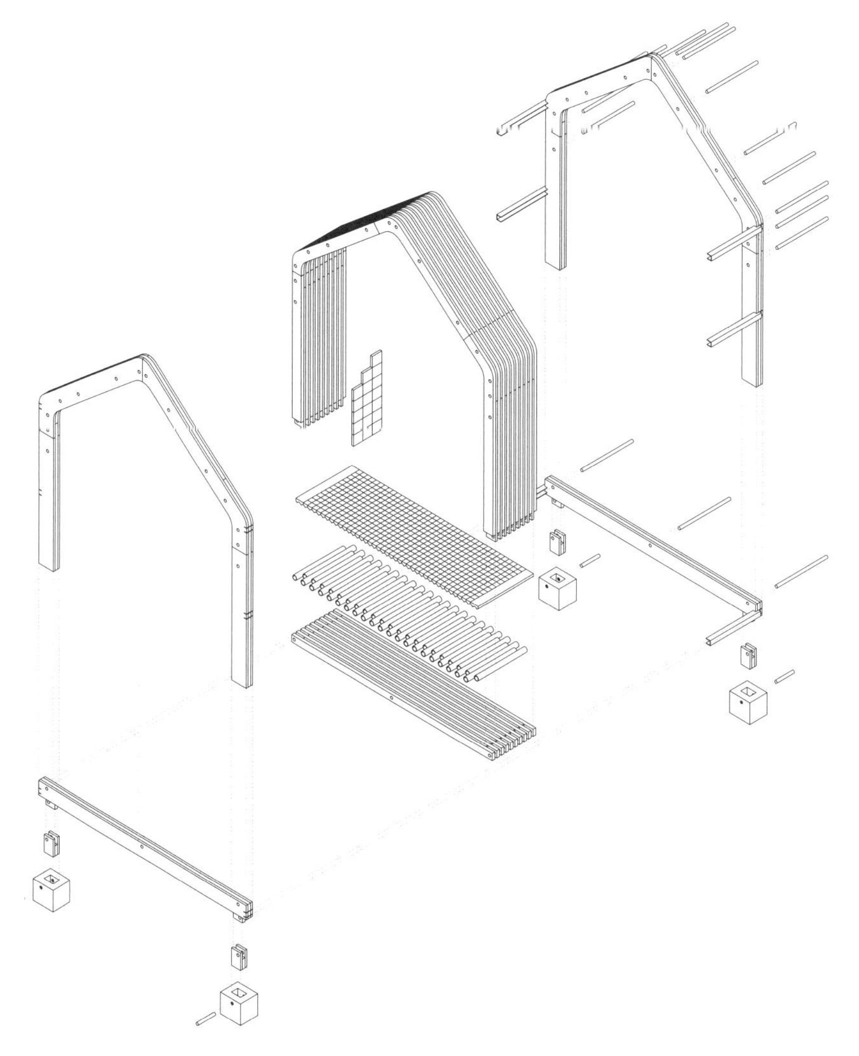

Exploded view of a BAMP! component segment of House 02 with detailed representation of the individual layers and connecting elements.

After the first experiences with paper-tube House 01 in terms of lightweight construction with linear components (frame construction), the focus in the follow-up project House 02 was on flat paper materials in the form of honeycomb boards or laminated corrugated sheets. This second prototype's basic design was comparable with that of the first BAMP! demonstrator, House 01 **» pp. 120–121.**

Thus, House 02 also follows the idea of a simple, rectangular building with a pitched roof. In addition to the choice of materials, the development of the optimal construction also involved taking into account the dead loads, wind loads and snow loads acting on the building. Structural integrity was then to be derived and calculated from the construction and the joining of the parts. The method of construction, especially the elementarisation and joining of the many individual parts, is based on the principle of layering honeycomb panels. The highest priority for the development of the structure was laid on the

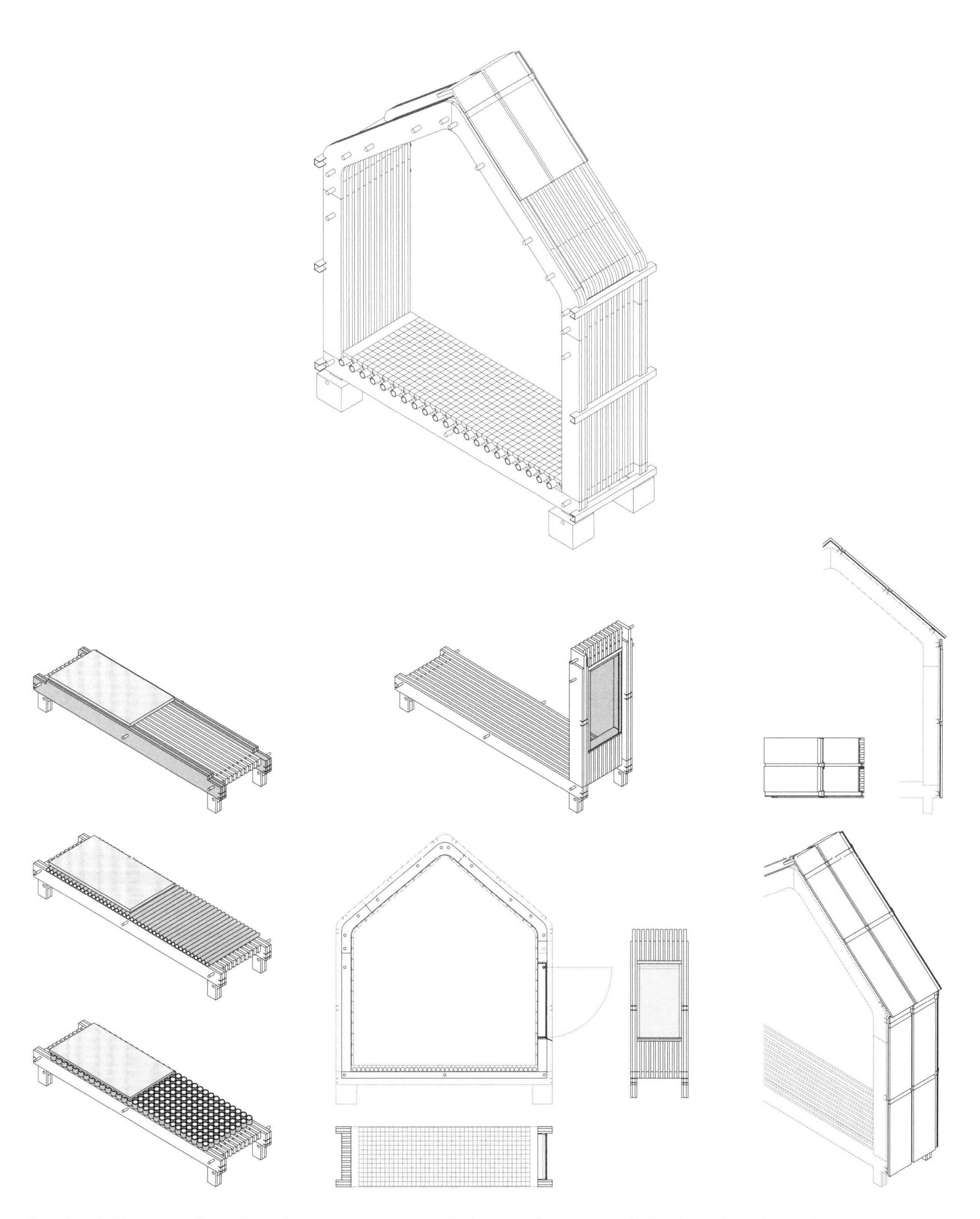

Planning BAMP! House 02: illustration of the structure of a segment (top), various floor structures (left and centre), section and floor plan (centre), outer skin as weather protection layer (right).

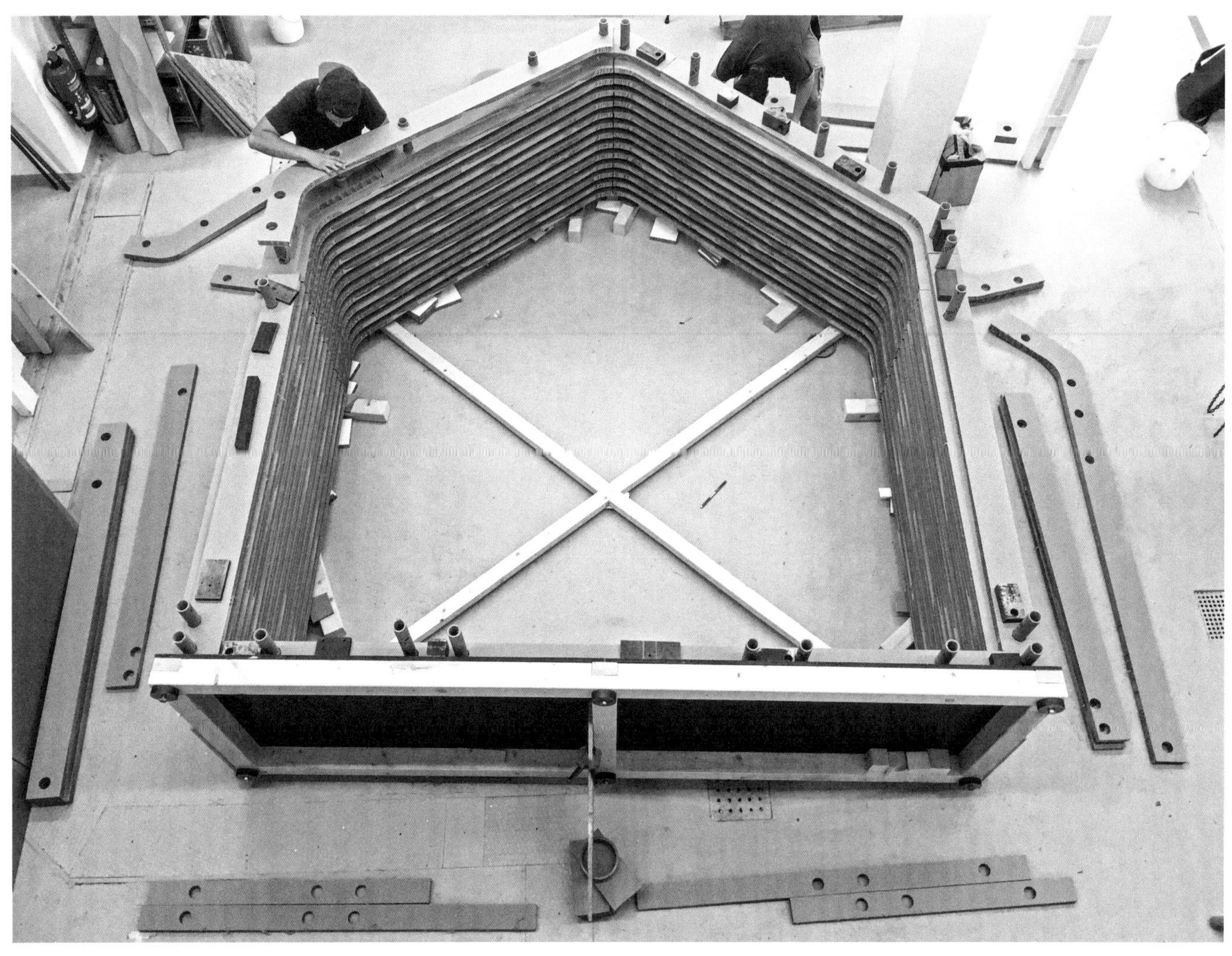

Construction of the demonstrator: here, the load-bearing structure is being built.

structural-constructive parameters.
The honeycomb panels were taken from large panels, with only a small amount of cut-off as the contours of the inner and outer sides of the honeycomb panels ran parallel.
The total wall thickness amounts to about 20cm, resulting in an almost monolithic layer with cut edges open in two directions. The walls and the roof layers were offset from each other by a few centimetres to optimise materials usage. The offset creates protrusions and recesses on both the inner and outer surfaces of the load-bearing frame elements. This makes it possible to install anchor points for potential inner and outer layers on the stable, flat side of the honeycomb panels.
The elements were cut from 2.38 × 1.6m panels using a CNC milling machine. The layer cross-section comprises 3 and 4cm thick panels. The individual elements were cut out and fitted with 5cm holes to accommodate the joining elements and pipe and cable ducts. Each frame consists of six individual parts. The joints of these frames are offset by approx. 50cm from one layer to the next to avoid continuous joints and weak points.
The pre-cut honeycomb core panel elements were glued together in layers. A stiffening wooden cross-frame, which served as an assembly aid and accommodated tolerances in the structure, kept the structure stable during, and made it easier to set up the segment after, assembly.
Assembly begins with the elements lying flat. The individual frame elements are connected with tubes that run through the milled holes. These tubes enable precise layering and also serve as the horizontal connection after erection.
The demonstrator was exhibited at the 2021 Venice Architecture Biennale as part of the "Building with Paper" installation.

Built demonstrator of the load-bearing frame structure of a BAMP! House 02 segment.

View of the “Building with Paper” installation at the Venice Biennale 2021.

The demonstrator at the Venice Biennale 2021.

EMERGENCY SHELTERS MADE OF PAPER

ARCHITECT/INVENTOR: Technical University of Darmstadt, Institute of Structural Mechanics and Design
LOCATION: Darmstadt, Germany
YEAR: 2020
USE: Emergency shelter or temporary housing
CONSTRUCTION TYPOLOGY: Frame construction
AREA: 16m²
PLANNED SERVICE LIFE: Temporary (18 months)

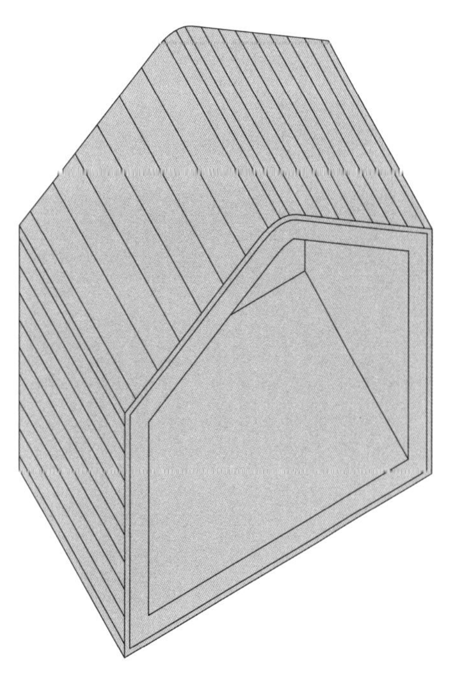

Axonometric view of the paper structure.

These emergency shelters made of paper are the result of research work by the BAMP! team and the paper construction research group of the Institute of Structural Mechanics and Design at TU Darmstadt. The aim of the emergency shelter project was to develop a modular construction method with commercially available paper materials that corresponded to the current state of the art. The pilot project is intended to demonstrate that paper constructions are very efficient and, precisely because of their simplicity, are suitable for large scale temporary housing construction.
In case of events such as earthquakes, floods and other natural disasters, there is a great need for temporary housing. Such units need to provide quick shelter for the occupants, must be easy to transport and rapid to set up. To conserve resources, it is also important that the building materials can be recycled. Due to its multi-layered construction, the shell of the emergency shelter fulfils all the required functions of a modern building. The aspects of transport, construction and jointing were kept as simple as possible so that even unskilled workers could assemble the emergency shelter on site and without complex tools.
In terms of architectural design, the emergency shelter reflects the archetype of a house, as the gable profile is understood as a pictorial depiction of living almost culture-independently. The modules are based on the 125cm grid typical of timber construction, and they can be combined to form different house sizes. They include various forms of housing with reduced space require-

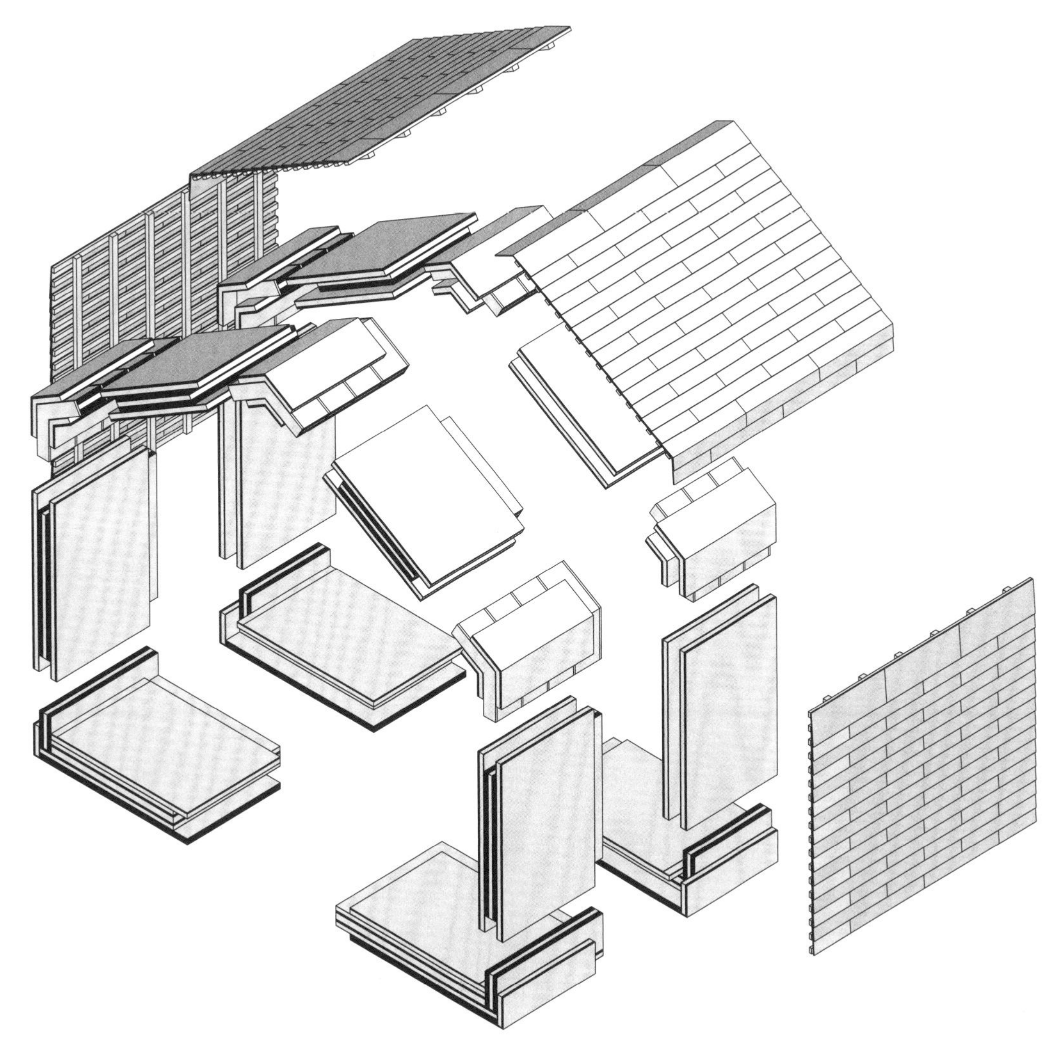

Arrangement of the construction segments to create the modules.

[illegible]

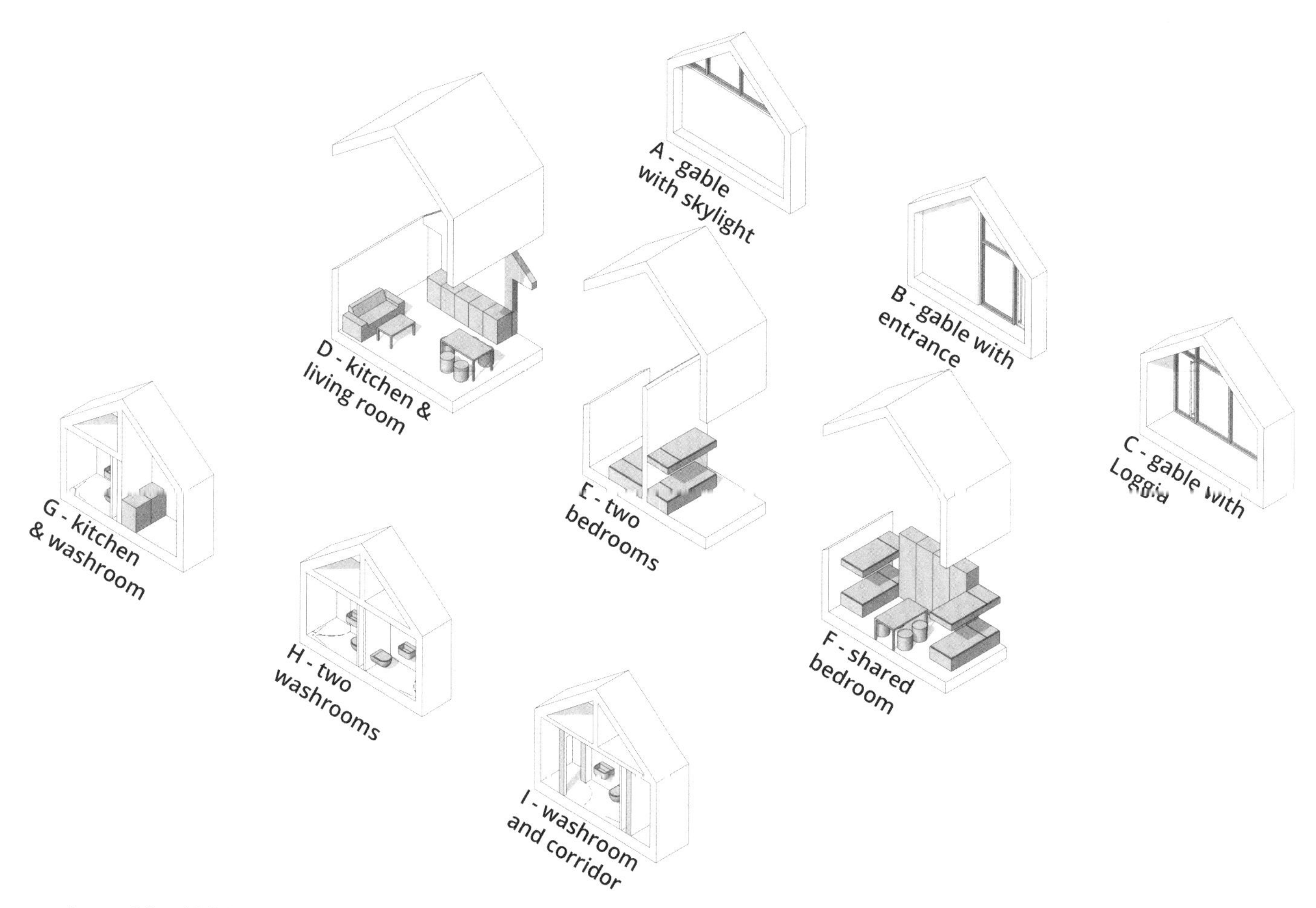

Various modules with living spaces that can be combined in different ways to provide an emergency shelter according to needs.

ments. There are end modules to close off the front ends, which also include the windows and the entrance doors to the residential units. The modular system additively serves the various living functions.

The solid paper structure consists of 25–30cm thick prefabricated segments built from multi-layer laminates. They can be combined with cardboards of different make-up.

The selection and arrangement of these boards depend largely on external influences and the functions to be fulfilled.

Exact material depends on the volume required, but the structure is predominantly made of corrugated board, which has great load-bearing capacity and whose small-cell structure insulates very well.

The three-layer structure of the wall segments is composed of an inner and an outer protective layer and a central core. While the protective layers shield the construction from environmental influences, the core layer primarily serves to stabilise the construction. It is the only one considered for structural calculations and in itself comprises several layers of corrugated and solid board and a honeycomb board core. The symmetrical protective layers also consist mainly of corrugated board, covered with fire-retardant solid board.

The decisive factors for the floor segments are the component's bending stiffness and the resistance to rising humidity. Unlike the walls, it is not corrugated boards but honeycomb boards that make up the largest part of the volume. Their cell structure, running orthogonal to the segment direction, ensures high compressive stiffness in the through-thickness direction of the panel. The water-repellent cardboard on the inside and outside of the respective segments reliably prevents moisture damage.

The prefabricated segments are connected to each other via a tongue-and-groove system. The middle layer takes over the function of the tongue, while the protective layers form the groove. This force fit

Structure of the wall core of the demonstrator modules: structurally effective core layer with inner and outer protective layers.

connection is strengthened or fixed by screw connections, similar to furniture connectors.
A rear-ventilated façade made of polythene-sealed cardboard shingles provides additional protection against the effects of weather, spray water, UV radiation and other environmental influences. The air circulating in the cavity allows any moisture to evaporate. In the event of damage, individual shingles can be replaced without great effort.

[illegible] weather protection layer made of polythene-sealed cardboard shingles

CARDBOARD THEATRE APELDOORN

ARCHITECT/INVENTOR:
Hans Ruijssenaars, ABT Building Technology Consultants
LOCATION: Apeldoorn, Netherlands
YEAR: 1992
USE: Theatre
CONSTRUCTION TYPOLOGY: Shell construction made of corrugated board
AREA: 240m²
PLANNED SERVICE LIFE:
Temporary (six weeks)

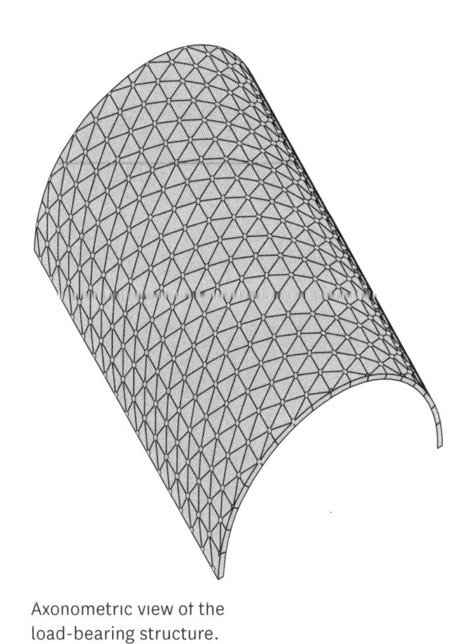

Axonometric view of the load-bearing structure.

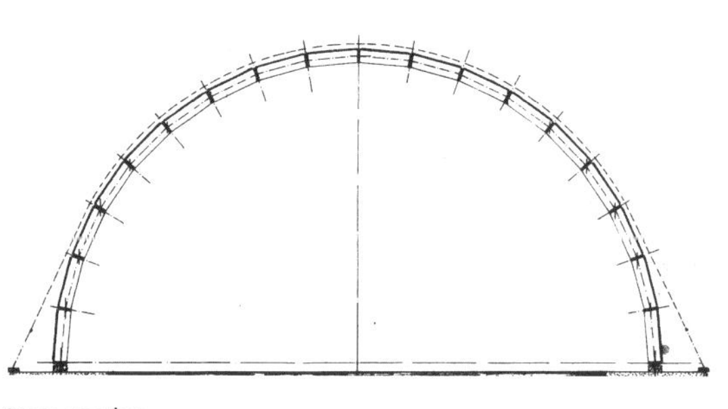

Cross-section.

On the occasion of the 1200th anniversary of the city of Apeldoorn, Prof. Hans Ruijssenaars from TU Eindhoven was commissioned to design a temporary theatre. The material of choice was cardboard, as Apeldoorn is close to the wooded Veluwe region where papermaking has a long tradition.
Ruijssenaars proposed a cylindrical design consisting of corrugated cardboard elements, all joined into a triangular grid and covered with triangular corrugated cardboard panels. This cylindrical shell spanned an area of 12 × 20.5m and could accommodate up to 200 visitors. The cardboard construction weighed only 1500kg. Each of the structural components consisted of seven layers of laminated corrugated board with additional hardboard inserts at both ends.
Six of the 1.2m × 350mm sized, and 35mm thick, elements were connected with a wooden disc and a hose clamp for each node.
The seams of the resulting grid of triangular corrugated cardboard panels were taped. All that was needed to assemble the shell construction was a hammer and a screwdriver.
The structure was mounted on prefabricated concrete slabs. A water-repellent fabric tarpaulin spanned the cylindrical shell; it was also anchored in the concrete planks and thus prevented the extremely light construction from being blown away.
The tarpaulin was attached to the top side of the nodes to avoid immediate contact with the cardboard panels and prevent them being wetted by condensation.
The Cardboard Theatre Apeldoorn was open to the public for six weeks, after which the construction was largely recycled. The project by Hans Ruijssenaars was an early example in which cardboard elements were used for a shell construction with a relatively large span.

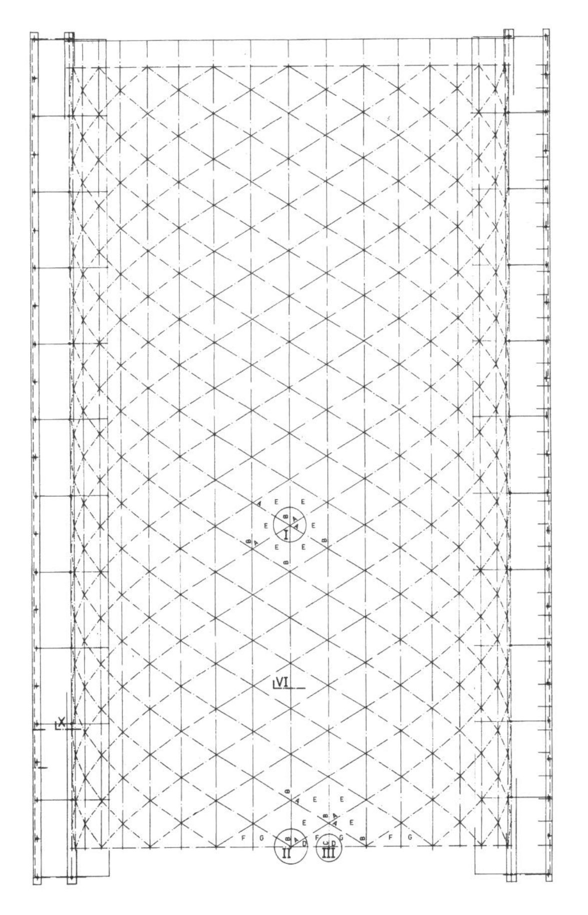

Roof plan.

The cardboard structure of the Cardboard Theatre Apeldoorn was covered with a fabric tarpaulin.

Interior view.

Detail of a connecting node

JAPANESE PAVILION (EXPO 2000)

ARCHITECT/INVENTOR: Shigeru Ban, Frei Otto, Buro Happold
LOCATION: Hanover, Germany
YEAR: 2000
USE: Exhibition pavilion
CONSTRUCTION TYPOLOGY: Shell construction made of cardboard tubes
AREA: 3090m²
PLANNED SERVICE LIFE: Temporary (five months)

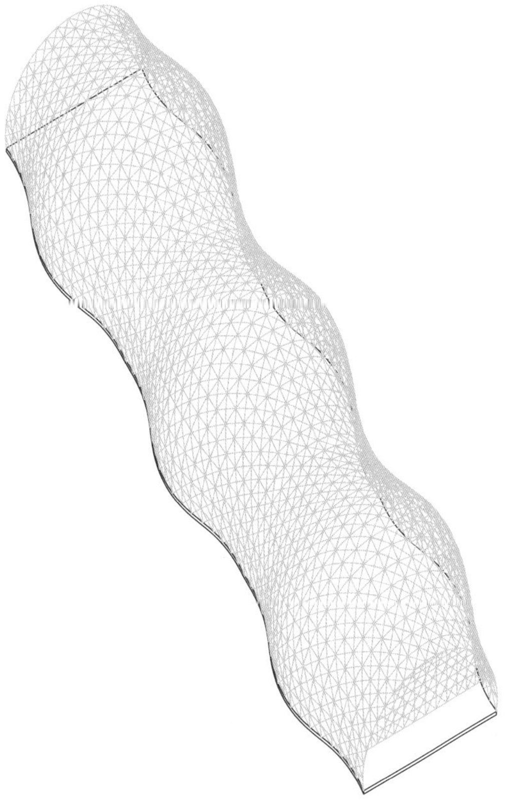

Axonometric view of the load-bearing structure.

Japanese Pavilion at night.

The theme of the Hanover Expo in 2000 was: "Humankind – Nature – Technology: a new world arising". In contrast to previous years, when the exhibitions aimed to showcase the latest technological achievements, the 2000 effort focused on the future aspects of humanity on the planet, following the goals and principles of the Agenda 21 action plan as defined at the 1992 Rio de Janeiro Earth Summit. The Expo opened on 1 June and attracted around 18 million visitors until the finale on 31 October 2000.
One of the projects on display was a pavilion made of cardboard tubes by Shigeru Ban. Ban had originally planned an arched shell structure similar to the Paper Arch Dome of 1998 or the Studio Shigeru Ban on the KUAD campus, **» pp. 112–113**. However, this would have required a large number of costly wooden connection elements.
Therefore, the decision was made in favour of an ecological and resource-saving construction, which was to consist largely of cardboard tubes and be almost completely recyclable. Instead of an arch construction, structural engineer Frei Otto developed a three-dimensional lattice construction made of long cardboard tubes that overlapped and produced a mesh width of 1m.
The final pavilion was 73.5m long, 25m wide and almost 16m high. At the time, it was the largest building made of paper-based components.
The plan was to assemble the 440 cardboard tubes, each 40m long and 120mm in diameter (wall thickness: 22mm), laid flat on the ground to form a grid, which was then to be raised with a scaffolding system. With this procedure, the shell achieved the desired double-curved shape and thus sufficient stiffness. Due to the limited transport possibilities, the 40m long cardboard tubes were divided into equal halves, connected with a wooden insert.

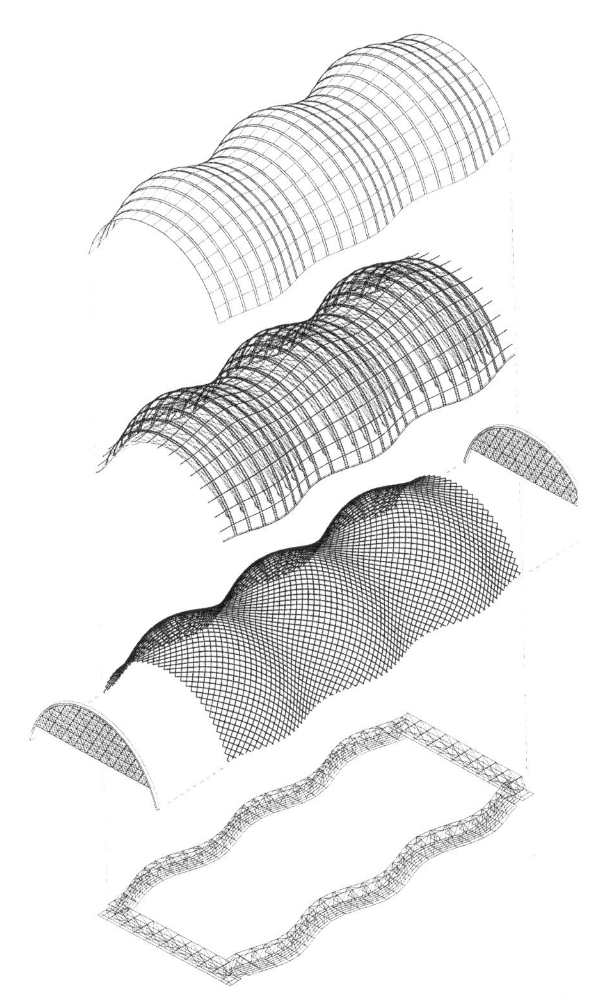

Exploded isometric, layers from bottom to top: steel frame foundation, layer of cardboard tube grid and gable walls, 3 × 3m wooden grid, outer membrane.

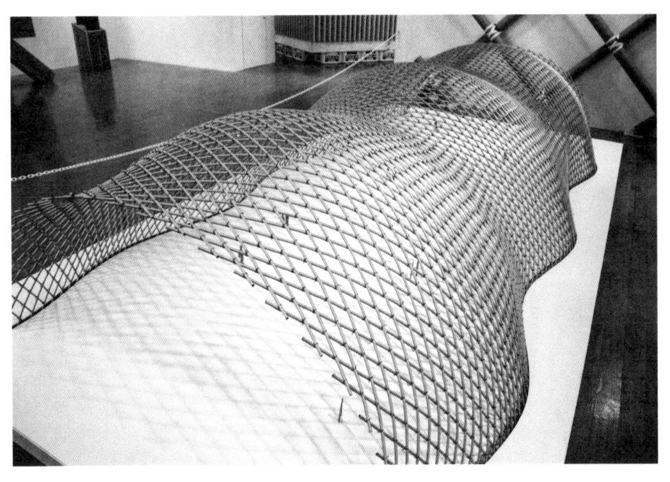

The model of the gridshell structure of the pavilion at the exhibition in Mito, 2013.

Connection detail in the gridshell model of the Japanese pavilion at the Mito exhibition, 2013. The fabric tape connection of the tubes in the realised building allowed three-dimensional movement of the tubes.

Tests showed that 120mm was the maximum possible diameter for the tubes to reach the required radius of 10m. Since cardboard tubes twist and take on a slight S-curve when compressed, the connection between the tubes had to allow for three-dimensional movement and rotation. This requirement was solved with a fabric tape.
Frei Otto suggested adding wooden ladders to the construction, running every 3m across the pavilion. The ladders were connected to each other by bent horizontal glulam laths, creating a wooden grid of 3 × 3m.
This grid served to fit the outer skin and facilitated construction and maintenance. Buro Happold added an 8mm slim stainless-steel cable bracing in the diagonal direction, which was fixed to the wooden grid and allowed free three-dimensional movement of the cardboard tubes.
A final coat of acrylic varnish protected the cardboard tubes from moisture. The inner membrane underneath the cardboard tubes consisted of five layers of flame-retardant polythene, non-combustible paper and a glass fibre fabric. For the outer skin, the designers chose a transparent, PVC-coated polyester fabric.
The cardboard tube gridshell rested on steel frame foundation boxes filled with sand. At each end, semi-circular wooden arches closed off the construction. The gable walls consisted of triangular plywood panels, honeycomb cardboard and a paper membrane that had been reinforced with diagonally running steel cables at a 60° angle to the foundation.
The Japanese pavilion was the largest building in the history of paper architecture at the time, and a milestone in its development.

PAPER THEATRE IJBURG

ARCHITECT/INVENTOR: Shigeru Ban Architects, Octatube
LOCATION: Amsterdam and Utrecht, Netherlands
YEAR: 2003
USE: Theatre
CONSTRUCTION TYPOLOGY:
Shell construction, dome made of cardboard tubes
AREA: 485m²
PLANNED SERVICE LIFE: Temporary

Top view of the paper dome.

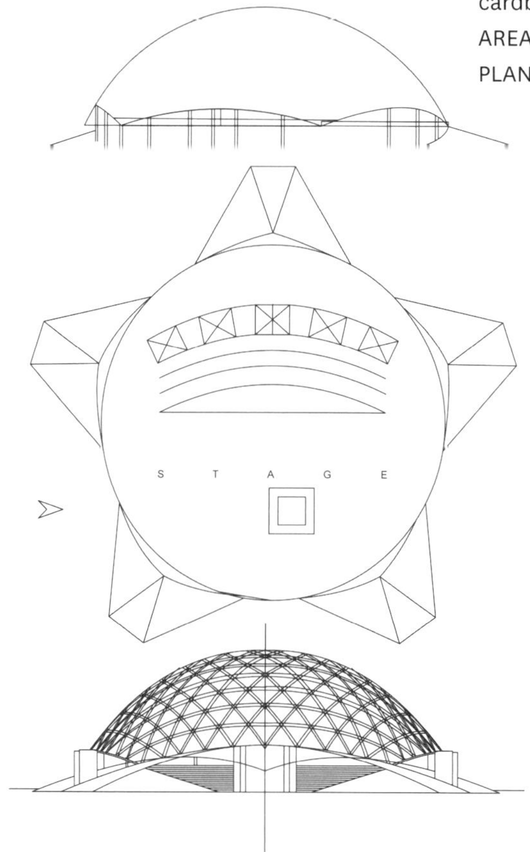

Elevation, schematic floor plan and section.

Jeannet van Steen, choreographer of the van Steen Theatre Group, asked Shigeru Ban to design a temporary theatre made of paper for her pantomime troupe. The theatre was to be located in the vicinity of IJburg, a newly developing residential area in Amsterdam, and later rebuilt in Leidsche Rijn near Utrecht. Since the dome was intended for a pantomime theatre, room acoustics did not play a decisive role.
Together with Octatube, Ban developed a geodesic icosahedron dome made of cardboard tubes. The imposing dome was 26m in diameter, 10m high and covered an area of 485m². The design idea is based on a traditional yurt – the tents of Central Asian nomads. About 700 cardboard tubes 1.2 to 1.5m long were used in the dome construction. The tubes' diameter was 190mm with a wall thickness of 18.5mm. In the course of the research and development for this project, it was found that cardboard tubes made from recycled material are about 40% weaker than those made from primary fibres. Each cardboard tube was closed with a steel muffle cover at either end. These muffle covers were connected with 10mm steel threaded rods running through the tubes. By turning the muffle covers, the rods could be tightened – thereby compressing and pre-stressing the cardboard tubes.
On the outside, the covers were bolted to star-shaped steel nodes consisting of six plates welded onto a steel pipe.
Five bent edge profiles made of IPE220 steel formed a tension ring at the lower end of the structure.
They were placed on five tetrahedrons that formed stable corner supports.

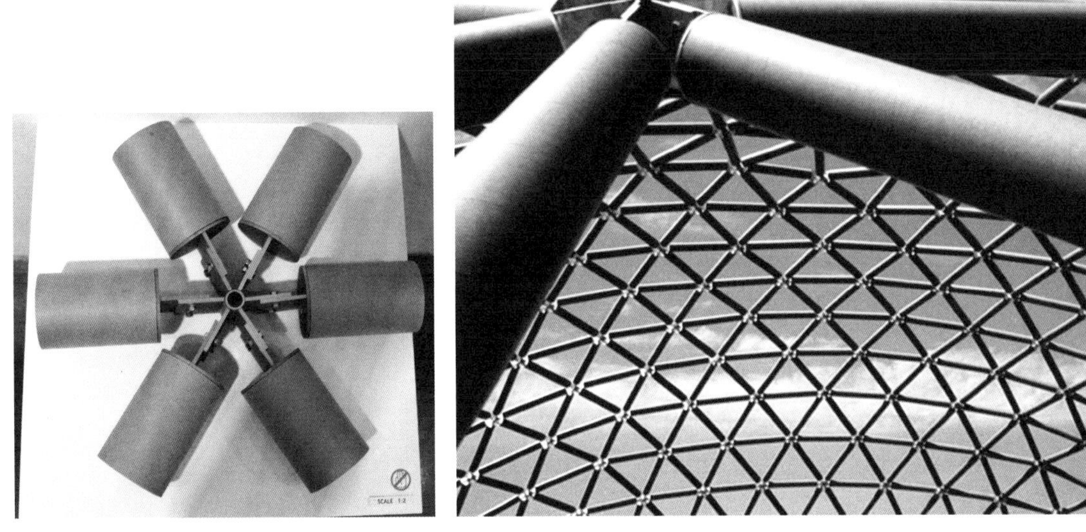

Steel node, top view, at the Shigeru Ban exhibition in Mito, 2013.

Shell construction made of cardboard tubes. The connection consisted of star-shaped steel nodes.

Steel node with spacer.

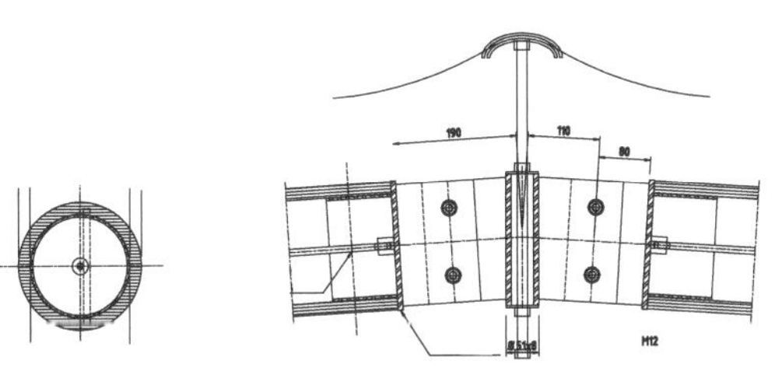

Steel node with spacer to the textile cover: plan view, cross-section and longitudinal section.

Since the highest point of the profiles' segmental arch was only 1.5m above the ground, the floor had to be lowered to allow comfortable walking in the space under the dome structure. In Utrecht, the designers solved this problem by placing the dome on an earth dike.
After the first row of cardboard tube triangles were mounted on the steel clamping ring and the dome was pre-assembled on the construction site, a crane took over further assembly.
The cardboard tubes were coated with [illegible] and 100mm inwards from the edges to protect them from moisture. A total of 663m² of PVC-coated polyester fabric covered the entire structure. The star-shaped nodes were fitted with mushroom spacers that could be pressed outwards and fixed in place to tension the fabric on top. Thus, there was no contact between the fabric and the tubes, which prevented condensation from forming and damaging the cardboard tubes.
The dome was used for various social events, accommodating 225 seated or [illegible] standing visitors. After being erected first in IJburg and then 2004 in Utrecht, the dome was dismantled in 2012 and put into storage. The construction was planned in such a way that it could be rebuilt at any time. Currently the re-assembly process is under way, and the building will be rebuilt in 2023 in Bijlmermeer, Amsterdam Zuidoost, to be used as a youth centre.

ARCH/BOX

ARCHITECT/INVENTOR: Jerzy Łątka (archi-tektura.eu), Agata Jasiołek, Weronika Abramczyk
LOCATION: Wrocław, Poland
YEAR: 2019
USE: Art installation
CONSTRUCTION TYPOLOGY: Column construction made of cardboard tubes
AREA: 38m²
PLANNED SERVICE LIFE: Temporary (one year)

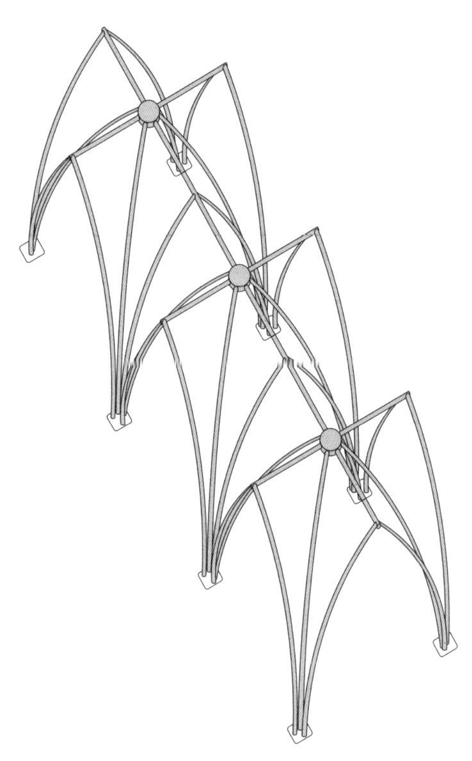

Axonometric view of the structural frame.

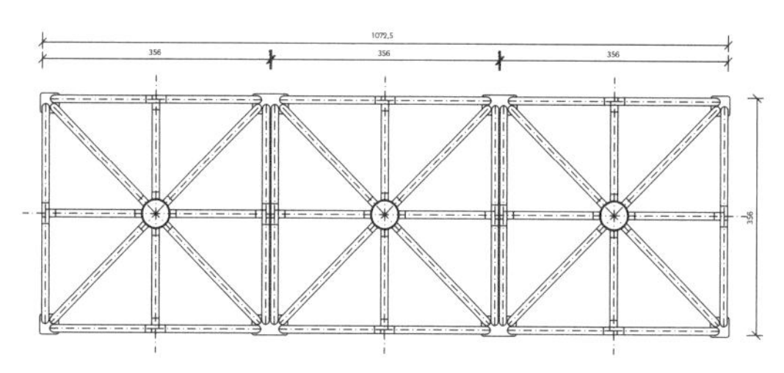

Cross-section and plan view.

ARCH/BOX was an experimental, temporary pavilion made of cardboard tubes. The architects drew inspiration for the design of the project from the Gothic style – an architectural era of which wonderful examples such as ancient cathedrals have existed for over 850 years. A construction made of cardboard tubes posed a counterpoint to the longevity we associate with Gothic buildings.

The load-bearing structure of the 10.7 × 3.5m-plan, 3.3m high, pavilion consists of cardboard tubes with an internal diameter of 100mm and a wall thickness of 7mm.
Wooden components connect the tubes; the "keystones" consist of a piece of cardboard tube with a diameter of 300mm and a wall thickness of 10mm. The curvature of the cardboard tubes was achieved by wet forming during the production process. This method allowed the creation of pointed arches that resemble Gothic forms.
The tubes were also slightly bent during assembly, which gave them an internal tension that additionally stiffened the construction.
Part of the experiment was to try out different impregnation processes. To this end, one segment was impregnated with an acrylic varnish while two other

ARCH/BOX in front of the Museum of Architecture in Wrocław.

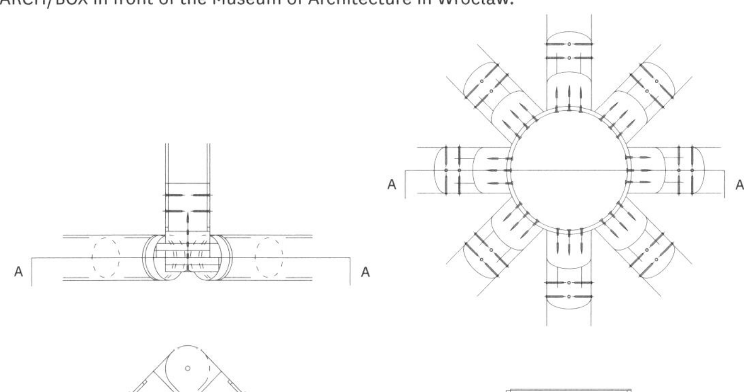

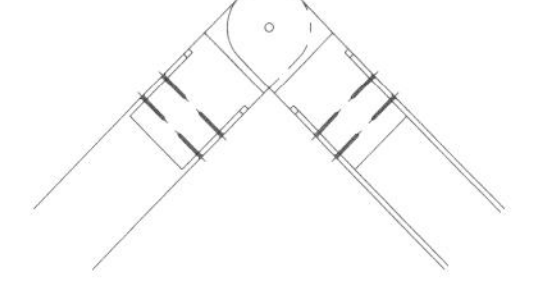

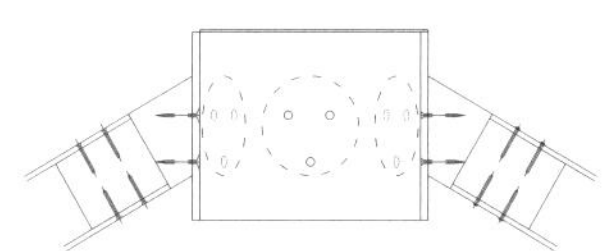

Detail of the front connection of the ARCH/BOX cardboard tubes; top view and section A–A.

Detail of the “keystone” construction: top view and section A–A.

Mounting the front connection.

segments were in addition to that impregnated with linseed oil varnish and wax.

After the pavilion had been exposed to wind and weather for seven months (even in winter), both impregnations showed the same protective properties. The project was developed and executed in collaboration with students from the Faculty of Architecture at Wroclaw University of Science and Technology – WUST

The front joints of two adjacent segments impregnated with different treatments

PAPER BRIDGE PONT DU GARD

ARCHITECT/INVENTOR: Shigeru Ban
CONSTRUCTION: Octatube
ASSEMBLY: Octatube with students of the Ecole Nationale d'Architecture de Montpellier
LOCATION: Vers-Pont-du-Gard, France
YEAR: 2007
USE: Bridge
CONSTRUCTION TYPOLOGY: Spatial truss construction
SPAN: 20m
PLANNED SERVICE LIFE: Temporary (two months)

Axonometric drawing.

Bottom view of the construction made of cardboard tubes joined together.

This paper bridge, realised by Mick Eekhout with Octatube and based on a design by Shigeru Ban, took up components that Octatube had previously used in the Paper Theatre in IJburg **» pp. 134–135** and a canopy in Avignon. The 2m wide footbridge was built near the Roman aqueduct Pont du Gard from the first century AD near Nîmes. The occasion was a cultural festival. The steps were also made of cardboard, obtained from waste paper, which was given a plastic coating. The construction comprised 280 cardboard sleeves with a diameter of 115mm and a wall thickness of 19mm, which were coated on both sides against moisture. The junctions consisted of galvanised steel, as in the two earlier projects. By integrating tension rods within the cardboard sleeves, the pre-stressed sleeves were assembled into truss arches. By means of the pre-stressing and the truss, asymmetric loads from pedestrians could be absorbed. The truss was then stabilised by diagonal tension cables. Similar to the theatre dome in IJburg, steel nodes connected the cardboard sleeves. Due to the pre-tensioning, a mechanical connection of the sleeves by means of screws in the nodes was not

Front view of the construction.

necessary as there was only compressive stress in the nodes. Due to the pre-tensioning of the steel rods inside the individual cardboard tubes, all tubes were subjected to compressive stress (developed by Luis Weber, Octatube). Thus, screw connections that might have torn out were avoided and, on the other hand, disassembly was very easy.

Bottom view.

Side view: The bridge spans 20m over the stream. In the background, the Roman aqueduct.

Assembly of the paper bridge [illegible]

PAPERBRIDGE

ARCHITECT/INVENTOR: Steve Messam Studio
LOCATION: Patterdale, Cumbria, UK
YEAR: 2015
USE: Bridge, art object
CONSTRUCTION TYPOLOGY: Arch
SPAN: 5m
PLANNED SERVICE LIFE: Temporary (ten days)

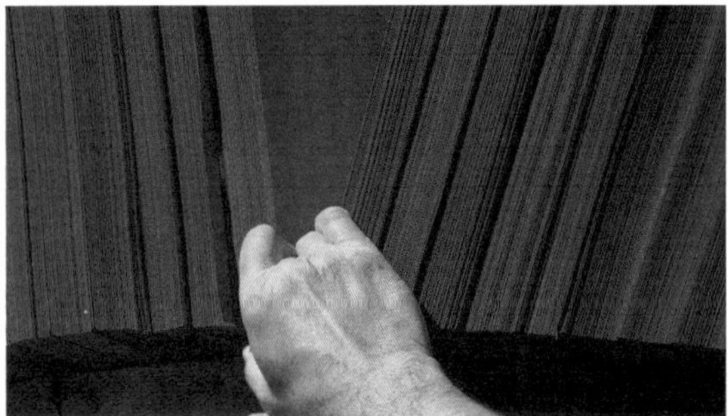

PaperBridge consists only of layered paper sheets without any fixings.

The bridge spans 5m across a stream.

PaperBridge is a temporary artwork made for and about the landscape of the Lake District (Client: Lakes Culture). On an aesthetic level, the piece acts as a focal point within a vast open landscape; the bright red colour draws the eye. The Lake District has a long history in the perception of landscape and aesthetics. PaperBridge is seen as part of its continuing legacy. Two footpaths travel up either side of the valley at that point, and a footbridge between the two thus serves a purpose. PaperBridge used an uncoated 270 g/m² paper specially made by James Cropper at the Burneside Mill. Cropper was able to produce paper of this particular vibrancy and colour that met the strict environmental standards this piece required. All the paper used in PaperBridge, 20,000 sheets altogether, was recovered and returned to Burneside

Detail view of the layered paper sheets.

View of the bridge with people walking across.

PaperBridge provides a focal point in the Lake District landscape.

Mill for recycling. PaperBridge uses no fixings but instead relies on the vernacular architectural principle of the arch. Its dimensions were 5 × 0.9 × 1.8m (length, width, height). An arched plywood form was placed between the two supports, enabling the paper blocks to be stacked in position across the river. A number of working models up to 1:4 scale were built prior to the construction to test the design concerning structural integrity and the impact of rain. Steve Messam also built a similar bridge in Suzhou in China in 2015 for Land Rover's 45th anniversary, which lasted only one day and was so stable that it could be driven on by a Range Rover.

A BRIDGE MADE OF PAPER

ARCHITECT/INVENTOR: Technical University of Darmstadt, Department of Steel Construction and Institute of Structural Mechanics and Design
LOCATION: Darmstadt, Germany
YEAR: 2019
USE: Bridge
CONSTRUCTION TYPOLOGY: Paper construction
SPAN: 6m
PLANNED SERVICE LIFE: Temporary (one day)

Bulkheads stiffen the bridge on the inside. In the same planes, reinforcements are placed from the outside between the upper and lower flanges of the I-beam.

A Bridge Made of Paper during load testing by Ulrich Knaack, Jörg Lange and Jens Schneider, Darmstadt 2019.

In 2019, interdisciplinary students designed a paper bridge as part of the IPBU project (Integral Project Building and Environment) of the Faculty of Civil Engineering at TU Darmstadt. The prefabricated construction of the "I-Beam Bridge" crossed the Darmbach in Darmstadt with a span of 5m and a width of 1m. The first stage of the project involved material evaluations and material tests.The results of these tests formed the basis for load-bearing capacity verifications. These verifications, in turn, were then checked and optimised using FEM (Finite Element Method) calculations.
According to the FEM analysis, the girder reaches a load capacity of 40 kN in terms of buckling and bulging. This analysis was verified by means of material testing. The upper flange represents the critical area of the girder, as most pressure loads occur in this area.
After a positive evaluation, the individual bridge elements were prefabricated. The elements were then transported to the botanical garden to be erected there.
For this bridge construction, paper girders (I-beams) were made from cardboard tubes, built as harnesses and

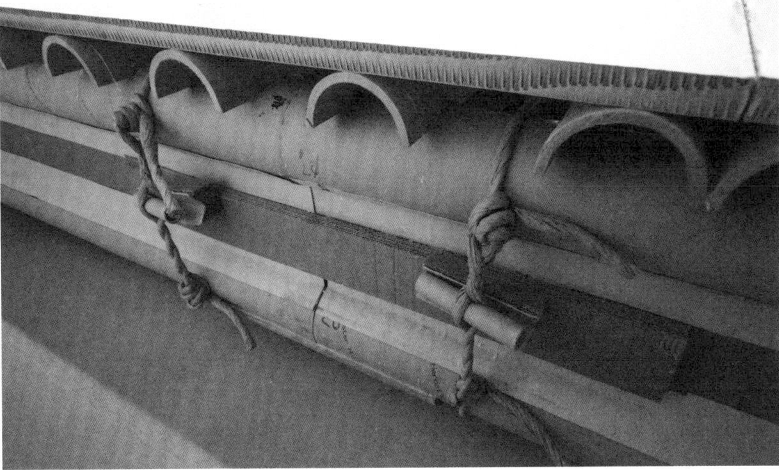

Detailed views of the bridge elements glued and temporarily fastened together.

Erecting the prefabricated bridge in the botanical garden.

The prefabricated top and bottom sides of the bridge.

solid board bars. The walking surface was formed by halved cardboard tubes and corrugated multi-wall corrugated boards. The only materials used for the construction were paper and glue; no additional materials such as screws or the like were employed. Weighing only 150kg, the bridge was capable of carrying over 500kg.

A total of three bridge elements were built, connected and glued together. Each element had I-beams on either side, connected by an inner tube. To stiffen the construction, four bulkheads were fixed between each of the longitudinal girders. In addition, solid board elements reinforced the upper and lower flanges of the I-beam from the outside.

AESOP DTLA

ARCHITECT: Brooks + Scarpa
CONSTRUCTION: RJC Builders
LOCATION: Los Angeles, USA
YEAR: 2014
USE: Retail
Area: 100m²
EXPECTED SERVICE LIFE: Permanent furnishing (estimated ten to 20 years)

The walls, the counter and the pendants are made of cardboard tubes. The worktop of the counter also consists of recycled paper.

Exterior view of the store.

Nestled in a ground-floor corner of the 1929 Eastern Columbia Building, Aesop Downtown Los Angeles (DTLA), a large store by the Australian cosmetics company, is situated in Los Angeles' historic theatre district, at the edge of its Fashion District and adjacent to the famous Orpheum Theatre.
Taking its cue from the neighbourhood culture and the cardboard rolls on which the bales of fabric from the nearby costume shops and fashion shops were wound up, the retail environment consists of round cardboard tubes of 15cm diameter that create walls and furniture. The vertical tubes are connected by screws. The pendants were also made from cardboard tubes. Countertops consist of recycled paper. For the display shelves, 25mm thick bamboo plywood was used. The three porcelain sinks for product testing were sourced from a local salvage yard. The floors are the natural concrete of the 1929 historic building, emphasising the vintage style. A long storefront with large windows creates a connection with the busy street.

View of the shelf wall made of cardboard tubes.

Interior view of the store.

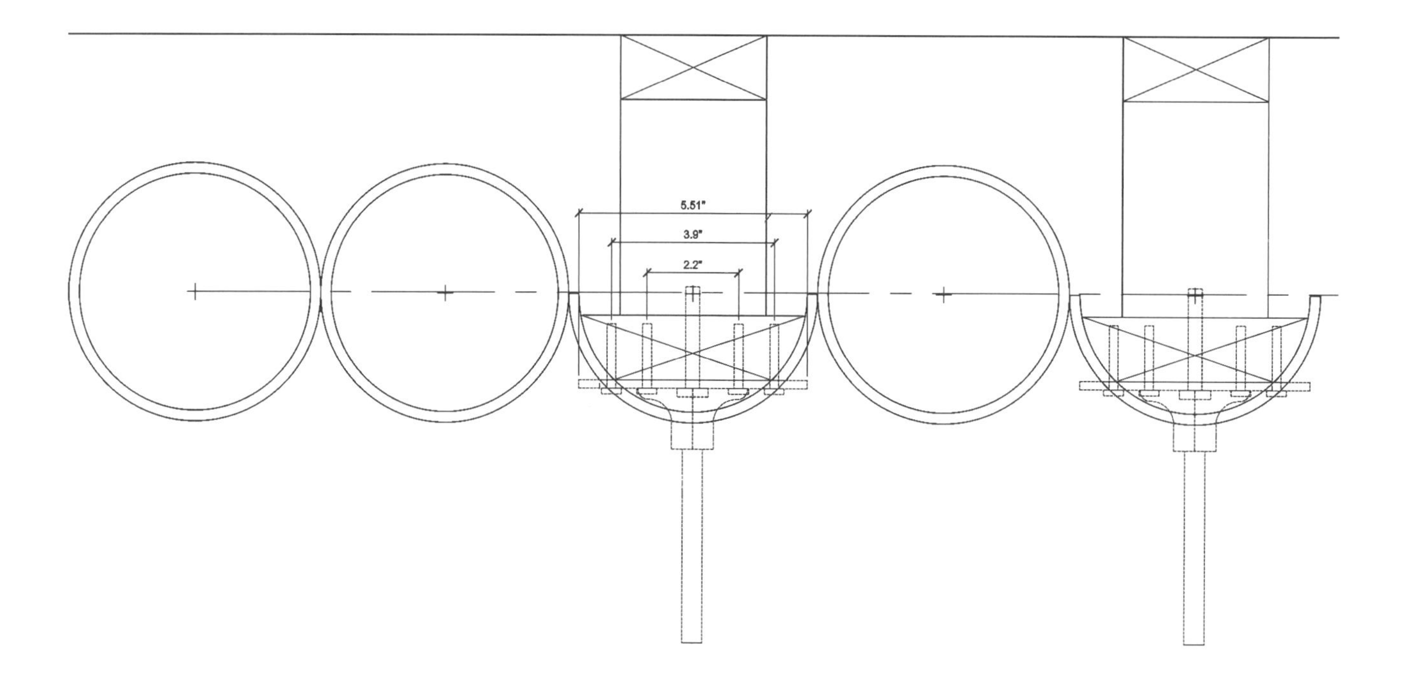

Horizontal section through the wall: to support the shelves, the cardboard tubes were cut in half and reinforced with wooden slats and screws.

Inlets in the wall for storage and presentation.

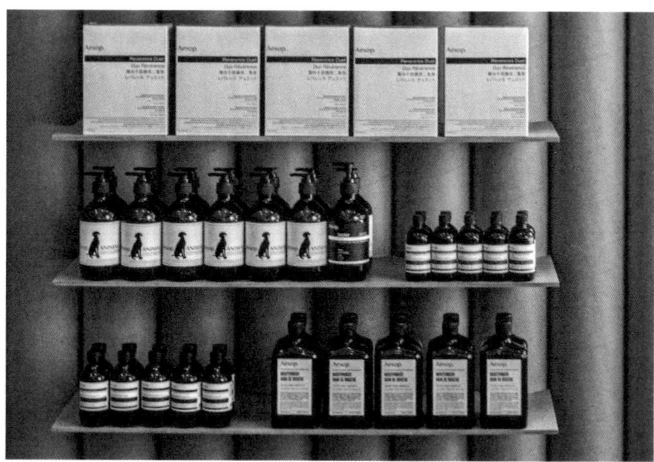

Detail view of a shelf wall.

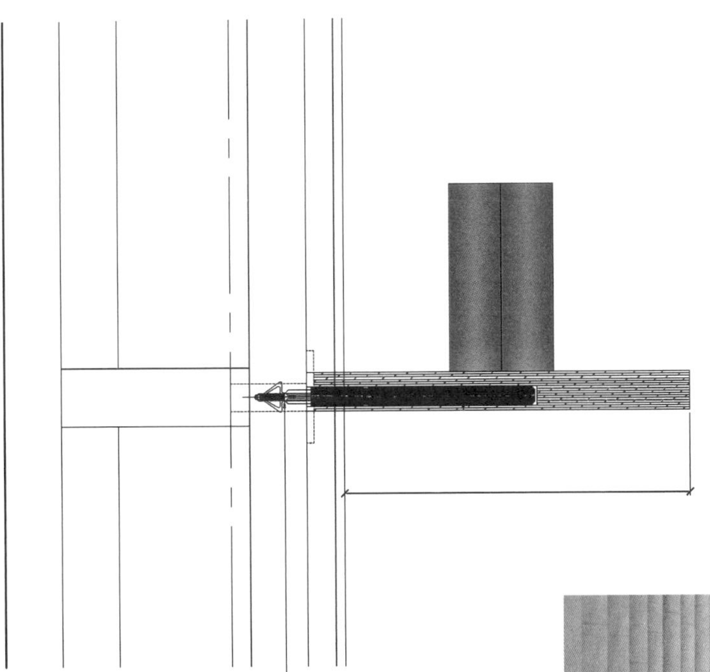

Vertical section showing the bamboo plywood shelves fastened in the cardboard tubes of the walls.

Overall view of the wall with cardboard tube cladding.

CARDBOARD BOMBAY

ARCHITECT: Nuru Karim, Nudes
LOCATION: Mumbai, India
YEAR: 2019
USE: Café
CONSTRUCTION TYPOLOGY:
AREA: 152m^2
EXPECTED SERVICE LIFE: Permanent furnishing (estimated ten to 20 years)

Exterior view: the corrugated cardboard structures are an eye-catcher for passers-by.

CardboardBombay is a café in Mumbai's central business district, Bandra Kurla Complex. The walls, chairs, tables and lamps were made from corrugated cardboard. Using a CNC router, curved, free-flowing shapes and textures were created from the flat corrugated material, which forms a kind of landscape with caves and protrusions. The architects' aim was to emphasise the versatility and durability of this recyclable material. Due to the packaging industry, corrugated cardboard is a widespread and thus inexpensive material that is usually made from recycled paper. Thus, the material has already gone through a complete life cycle and can undergo various further cycles if disposed of correctly. The combination of corrugated layers with cover layers gives the material a high degree of stability with a low weight. The architects took their cue from the wavey structure of the corrugated layer and transferred it to the large curved form language. Tables and benches seem to spring from the cardboard landscape. The different cutting angles create

Interior view of the curved corrugated cardboard landscapes.

varying distances between the individual waves of the corrugated board, which results in a play with the perception of perspectives. Prior to construction, the designers built prototypes and the material was tested for water resistance and temperature fluctuations. The table tops were impregnated with wax to protect them from moisture and facilitate maintenance. The open surfaces also absorb sound and create diffuse reflections, significantly reducing the noise level and thus achieving a more private atmosphere. The ground-level café is 152m² in size and can accommodate over 40 guests. The production of the cardboard elements was handled by Harosh [illegible] Packaging.

Both the chairs and the tables are made of corrugated cardboard.

The curved wall structures were made with a CNC milling machine.

The different cutting angles create a playful, perspective distortion of [illegible]

Table and bench structures seem to grow out of the corrugated cardboard landscapes. Due to the high stability of the corrugated board, large cantilevers are possible.

The table surface was impregnated with wax as a protection against soiling and damage caused by moisture.

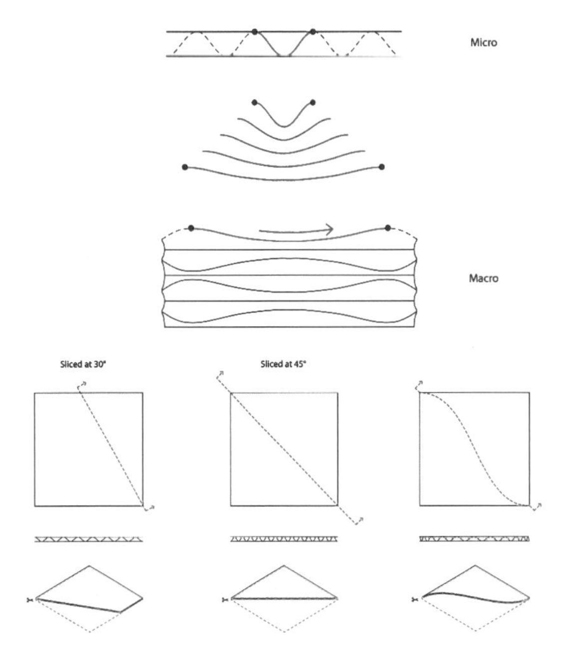

Design sketch: wave structures in the micro (wave layers) are transferred to curved structures in the macro, i.e. on walls and furniture. Different cutting angles create an optical play with the [illegible]

CARDBOARD OFFICE PUNE

ARCHITECT: studio_VDGA
LOCATION: Pune Pimpri-Chinchwad, India
YEAR: 2020
USE: Office
AREA: 1254m²
EXPECTED SERVICE LIFE: Permanent furnishing (estimated ten to 20 years)

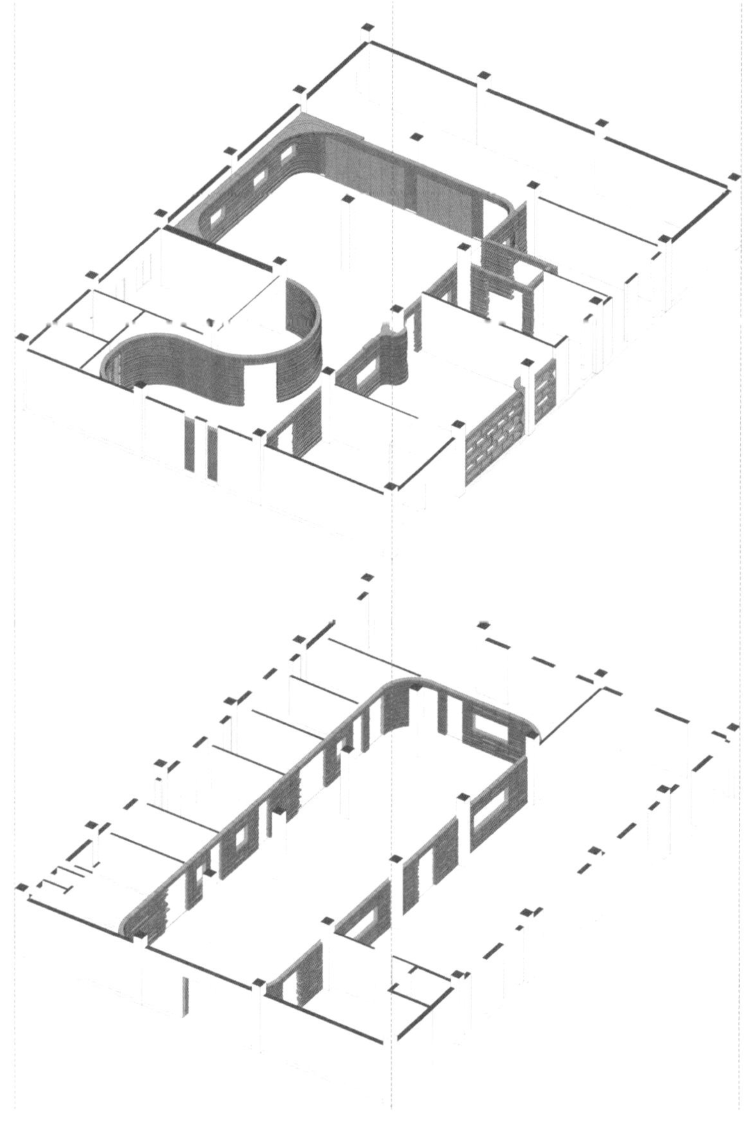

Axonometric drawings of the office space on the ground and the first floor.

Interior view.

In this office space, comprising four levels, the ubiquitous solid partitions were replaced with honeycombed cardboard – thus adding texture and form to the space. The honeycomb boards were laid in layers in different profiles so as to form free-flowing curves. Honeycomb board is an inexpensive product. Its hexagonal inner cellular structure is endowed with good compression strength and rigidity. Owing to the thickness of the board, it provides sound absorption as well. The boards were also used for the reception area and the doors.
The workspace serves a manufacturer of electronic components with a staff of almost 100 people, providing them with an unconventional office space. The free-flowing cardboard walls create textured partitions that envelop the work stations, conference zones and lounges. Curving cardboard elements create wall sections that billow into the room or wrap around columns. In some areas, the cardboard sheets' exterior has been cut away to reveal the internal honeycomb in

Interior view with workspace.

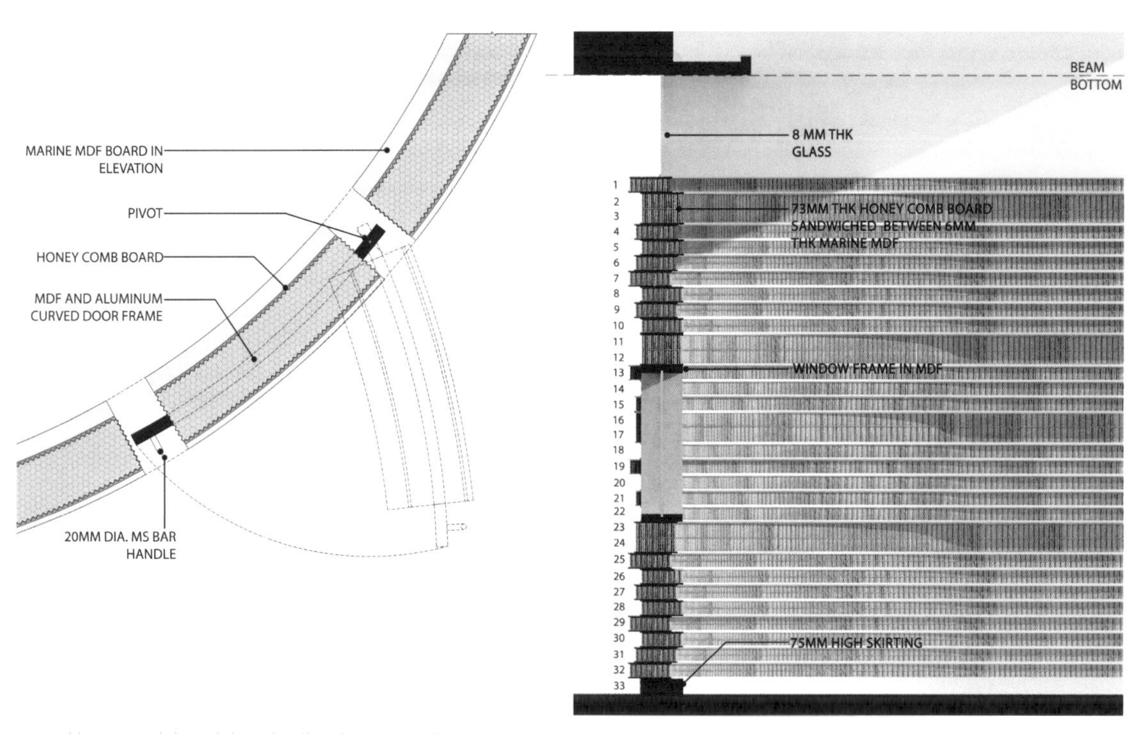

Curved honeycomb board door detail and typical wall section

Detail of a curved cardboard wall with open end.

View of a window and the integrated services.

order to create an interesting texture. Transverse cuts through the nodes of the hexagon reveals sharper fins, whereas longitudinal cuts through the board exposes uneven wider bands, resulting in a play of shadows.
Slim horizontal slots form windows. An installation of plants and electrical components in white paint tins left over from the refurbishment is suspended from the ceiling. The raw ceiling with exposed electrical and AC ducts was conceived as a part of the interiors in order to maintain the height of the space. While the lower three floors comprise workspaces, the terrace floor was converted into an enclosed dining area and a semi-open canteen. The time-consuming process of cardboard cutting and installing it layer by layer was done on site.

Interior of a conference space.

View of a large office space.

Entrance area with art installation.

CARTA COLLECTION

ARCHITECT: Shigeru Ban
PRODUCTION: wb form Zürich
YEAR: 1998
USE: Furniture

Paper chairs in a corridor of the Miyake Design Studio Gallery.

Detail of the chairs: The seats are made of cardboard tubes, which are attached to frames made of birch plywood.

For the Miyake Design Studio Gallery in Shibuya, Tokyo, as well as for his own home in Japan, Shigeru Ban designed the Carta Collection paper furniture series in 1998. It comprises seven elements: a stool, a chair, a large chair, a bench, a chaise longue, a round table and a coffee table consisting of cardboard tubes supporting a white glass top. The seats are made of thin cardboard tubes of recycled material strung together, supported by frames or legs of European birch plywood. The cardboard tubes are sealed with a urethane resin to protect them from moisture penetration. The chaise longue measures [illegible] × 57 × 202cm, the lounge chair is 70.4 × 57 × 71cm and the bench is 44 × 57 × 205cm. The shape formed by the cardboard tubes of the table, which measures 37.5 × 88.5 × 55cm, is reminiscent of the curved structure of the Paper House by Shigeru Ban » **pp. 86–87**.

Table construction made of cardboard tubes with glass top.

View of a seating group consisting of chairs, stool and chaise longue.

7 OUTLOOK

Why build with paper at all? The book opened with this question, and it concludes with this question. The previous chapters have described the current possibilities of using paper to construct buildings. There is no question about it; this building material has a lot of potential:

- it is environmentally friendly because it is made from renewable raw materials;
- it meets the requirements of the circular economy because it is recyclable;
- it has good insulating properties; and
- it has excellent specific load-bearing capacities.

Paper materials can be produced in numerous shapes and forms, thus taking on diverse functions and forms of construction and design. On the other hand, there are significant challenges to overcome when building with paper – especially with regard to fire protection, moisture protection and durability. The chapters in this book have shown that there are solutions to these issues.

But generally speaking, the development of paper as a building material is still in its infancy. So far, neither standards nor technical rules exist. But what we do have is qualitative research results and initial empirical data from existing paper buildings » **fig. 1**. This chapter will focus on how designers, architects, constructors, civil engineers and paper machinery engineers assess future opportunities and challenges from their respective perspectives.

Redefining design

What does architecture look like when it is made of a specific material? What characterises paper architecture? Which aesthetic properties does a construction, a building or a piece of furniture have if it is made of paper? How do the design and manufacturing processes affect form and shape? Which constructive necessities must the design of a paper building address?

Paper as a new building material poses a problem for the established construction processes. Experience is limited, and buildings made primarily of paper usually come in the form of prototypes and special constructions and are few and far between. So how does a shape, a design, an architecture develop out of the material? The first obvious steps in working with paper and cardboard are to transform processing methods and joints from other materials: in other words, to adopt principles.

Standard building components and elements made of paper have yet to be developed. If neither the material in its final form and surface quality nor material-specific construction and processing methods exist yet, then not only does the technology need to be developed but shape and form need to be developed and defined as well. Haptic and optical properties are associated with every material, giving it its appearance and character.

So, which special processing methods for paper and cardboard have a formative effect? Is it the folding method that is specific to paper-based construction, or is it the block-shaped element consisting of several layers from which a specific shape or part is milled? No, it is the construction methods that result in very specific building structures and generate their form.

The example of a beam made of honeycomb boards can be understood as a "cardboard model" of a double T-beam »**fig. 2**. Obviously, adopting principles – be it the shape or the dimensions of a building element – does not automatically result in the same properties. However, this transformative process drives the designer to make initial comparisons with existing materials and to rethink the familiar. Copying the form and function of a steel double T-beam with paper materials leads to something new: a similarly shaped component with material-specific aesthetics. Only the processing, the joining and the use in architecture or furniture construction give the material its form and thus its final shape.

For materials such as wood, stone and metal, material-associated structures have developed over a long period of time. In the case of the building material wood, for example, these range from historic half-timbered houses to modern wooden buildings.

1 Model of a paper wall structure: BAMP! House 02, TU Darmstadt, 2019.

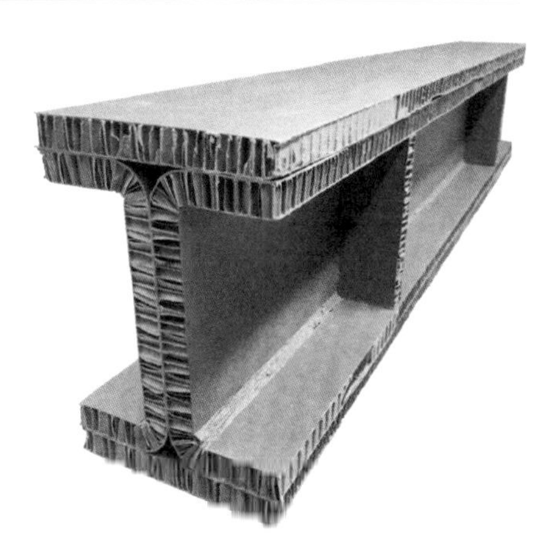

2 Cardboard double T-beams.

3 Paper shingles in analogy to bitumen or wooden shingles.

Traditional building methods have been adapted and optimised over time through experience and technological development.

Construction methods such as half-timbering led to a structural aesthetic that associated the appearance of the architecture of a particular time with a particular material. Since new findings change the shape of buildings, timber construction – to stay with the example – has developed from timber framing and log houses to a systematised construction method with panels and elements. A processing method defined by joint type and state of the art evolved into a new way of building with wood through industrial production and the layering and laminating of the material. This process is associated with new design options for designers and contractors. New construction methods allow buildings to have different, more flexible floor plans and enable alternative façade designs. The concept of replacing familiar prefabricated concrete building systems with prefabricated timber construction opens up new architectural horizons. Design-defining construction methods, such as cantilevered fair-faced concrete slabs – which contradict today's thermal insulation requirements – are now conceivable in different forms with a similar appearance. Composite or hybrid construction methods let the recognisable materiality of a construction method and its material-typical shape fade away. For some building materials, our material knowledge lets us derive certain design principles, which are assignable to timber, steel and masonry construction. Thus, the material is linked to a prevailing look and aesthetic. Does this mean that a new material necessarily leads to new construction methods?

And, with this in mind, how do new mechanical or manual processes affect the aesthetics of a building? Taking the simple example of paper shingles for a façade cladding: one can choose from many common products made of standard materials such as bitumen or wood. Transferring the shape of these products to a paper product is the

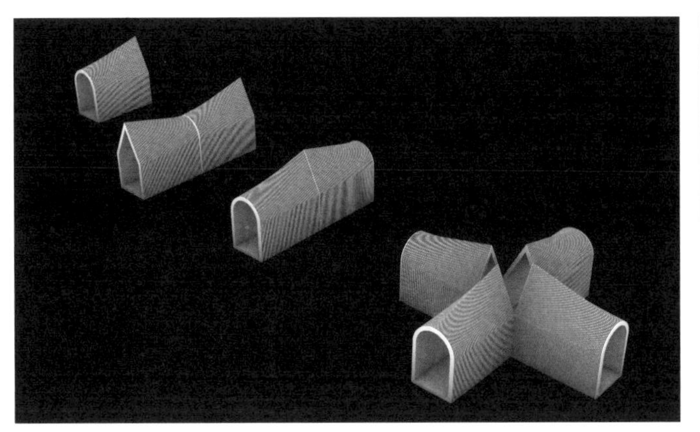
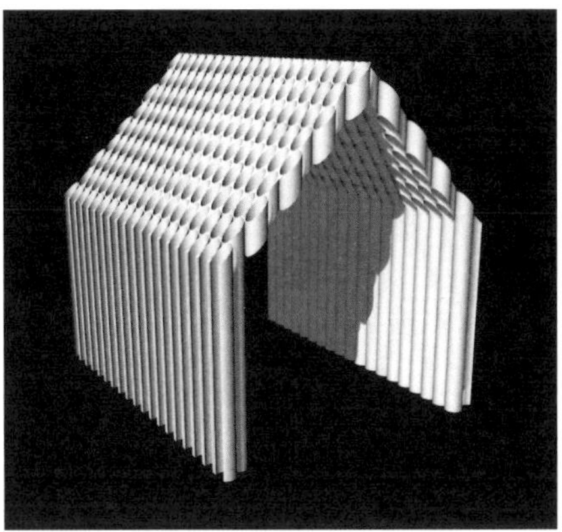

4 Designs of spaces created with paper materials.

first step – but understanding the material's potential and translating it into a new form, a new structure, stems from the material itself and becomes formative »**fig. 3**.

A paper construction method in the sense of a material-specific canon of its own does not yet exist. But in the process of building and designing with paper and cardboard, specific solutions and shapes are created for the respective requirements. One day, the multiplication of attempts and possibilities might generate such a construction method. Therefore, everybody actively involved in construction is called upon to define and further develop the design for paper buildings through their actions.

Function and use

In addition to the lack of a clearly defined design language for paper constructions, the functions that can be applied to paper-based buildings or components still allow for a broad development field as well. At the moment, paper as a building material does not promise a 100-year lifespan for any kind of building – meaning that, for the time being, paper cannot be used to construct durable structures.

Consequently, the issue is about recognising, structurally investigating and applying those functions that paper as a building material can perform. Paper often performs better than, for example, a textile: it offers more mass and, in the form of corrugated board, features sufficient insulation properties. Paper is virtually unbeatable in terms of ecological aspects: it consists of renewable raw material and, when used appropriately, it is easily recyclable. It is theoretically indefinitely available and usable in unlimited quantities. Accordingly, temporary buildings that are intended to be used for longer than just a few weeks but that do have a defined end-of-use date are predestined to be made primarily of paper. For such uses, paper material offers a good alternative to solid and mineral-based buildings.

It is the designer's task to find areas of application for paper and cardboard »**fig. 4**. – but also to draw boundaries where they are unsuitable and should be replaced or combined with other materials with more appropriate properties. In lightweight or interior construction, for example, paper's good material recyclability and possibly the simpler deconstruction steps could create an alternative to conventional lightweight plasterboard construction, opening up new potential. Factors such as

5 Design of the emergency shelter "Cardboard Container House", flexible living in a paper house – TU Darmstadt.

sound insulation, fire protection, thermal insulation, structural properties and visual properties must be defined and tested in this respect. There are many buildings and lightweight solutions that are short-lived. One example is trade fair booth construction: structures that are designed to last only a short time and are usually disposed of as hazardous waste well before their life expectancy runs out. Structures made of paper and cardboard could therefore replace existing systems in lightweight interior construction. One advantage would be the simple and ecological dismantling. To introduce such change, the utilisation times of specific constructions must be defined, and the usability of paper materials and cardboard must be optimised.

Processes and concepts in the building industry are usually designed for a long service life. Consequently, building construction, design and function are also subject to the same objective. A less durable but more easily dismantlable and recyclable building material offers new options for architecture and construction. Paper is predestined for constructing buildings with shorter use and lifetime in the knowledge that disposal or modification poses no problems in terms of ecological and sustainability aspects »**fig. 5**. Building construction could become easier again because the material "de-technologises" the construction process. The idea of a simple cardboard hut that is home-built and whose repair or disposal can be done by the user – a strange idea or bold vision? The need for temporary building solutions, changeable architecture and changeable spaces has long existed.

Construction

Paper as a building material is still new, and experience with its use in the construction industry is correspondingly limited. Consequently, there is considerable uncertainty in the constructive handling of the material, which is why it has hardly been used so far. Structures made of paper cannot be planned and built according to known sets of rules, and the associated uncertainties lead to either choosing other materials or to longer design phases, larger safety margins and thus higher costs. In order to raise the potential of using paper as a building material, one has to deal with and use it regularly, starting with small parts and building components, if necessary. This is the only way to improve the understanding of the material and its performance.

In a building, the interior design and construction finishing are ideal for experimenting with paper as a building material because of the relatively low moisture-related risks and low loads experienced here. And this is where the advantage of the easy recyclability of paper comes into play, as the useful life of interior construction finishing is ten to 15 years, which is significantly shorter than that of the entire building.

Building envelopes without a load-bearing function represent another interesting area for using paper as a building material if individual components can be exchanged easily. Such components only have to bear their own weight and the live loads per floor. The heat-insulating properties of corrugated or honeycomb boards are advantageous here. In order to avoid moisture damage, however, structural moisture protection is vital.

Although most experimental constructions are found in load-bearing structures, they always have to deal with the problems of moisture penetration and fire protection. Coatings interfere with recyclability, and replacing damaged components is usually difficult. However, it is noticeable that most such structures use skeleton rather than solid constructions, even though the latter can more easily meet fire protection requirements. This, therefore, poses a field for further development.

Progress can certainly be expected with paper products that build on the development of new materials. Paper products that specifically allow for force application and force concentration will enable slimmer and thus more efficient designs. The same applies to recyclable coatings that protect the paper material from moisture and fire.

Another area where research suggests potential relates to the introduction of additional functions such as electrical conductivity, the possibility of controlled deformation or the dissolution of parts of the structure under certain definable circumstances (humidity, temperature, load).

6 Textile connection (Tanzbrunnen Cologne) as a model to further develop the joining of thin-walled paper constructions.

As far as joining technology is concerned, building with paper is mainly oriented on timber construction and adapts corresponding systems and components. The adapted solutions are promising, as they are most appropriate for specific properties of paper such as anisotropy. In addition to mechanical connections, bonding and adhesive technologies are of great importance in building with paper. They have been very successfully advanced in timber construction in recent years. Bonding allows forces to be distributed across a wide area of the material, which is particularly advantageous for paper structures.

Joining techniques from the textile sector also lend themselves well to building with paper: since it is a flat material, textile can better absorb tensile than compressive forces »**fig. 6**. Corresponding solutions with seams, welting connections or bonding are easily conceivable and have development potential. Finally, the aspect of additive manufacturing should be pointed out. Volume-generating manufacturing processes that produce free-formed components from paper mass are conceivable, and initial additive manufacturing attempts are promising »**fig. 7**.

Material technology

One of the reasons for the project "BAMP! – Building with Paper" was the fact that up to now, all buildings made of paper have been made of materials such as corrugated board that are actually intended for a completely different industry, namely the packaging industry, the only exception being cardboard for plasterboard. There has been very little exchange and hardly any specific product developments at the interface between paper manufacturing and the construction industry. The BAMP! project contributed to forming corresponding networks at this interface and promoting an understanding of the mutual requirements and possibilities. This also gave rise to new ideas and perspectives for developing new applications for building with paper.

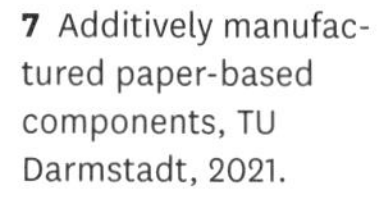

7 Additively manufactured paper-based components, TU Darmstadt, 2021.

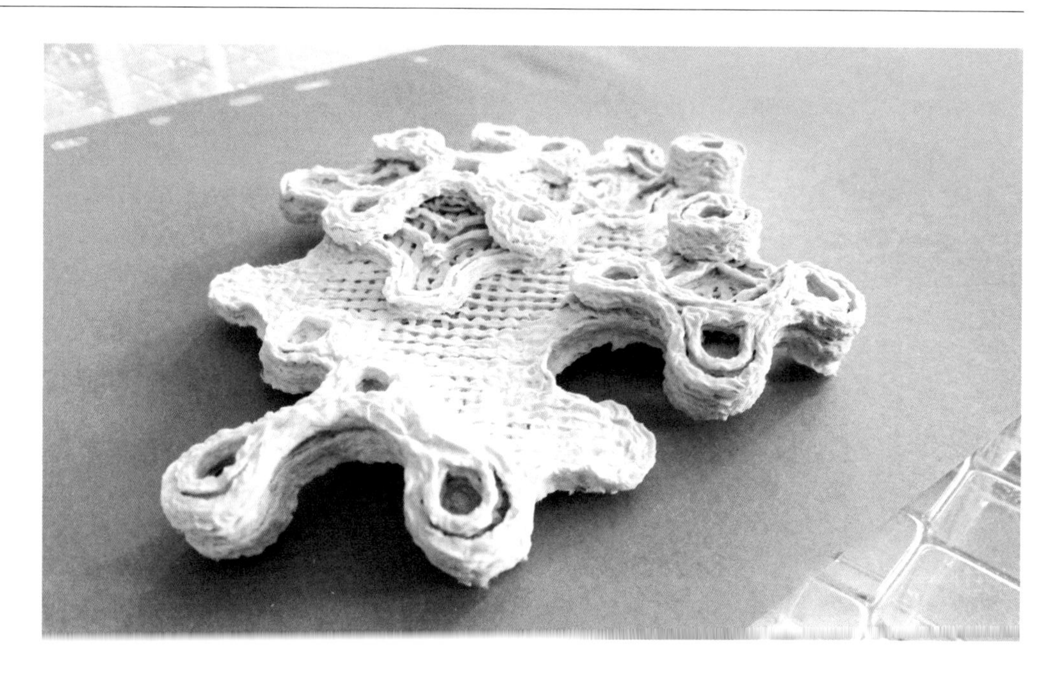

8 Transparency variations in paper-glass laminate.

10 Paper web with a weight component of 1% recycled carbon fibre. Extraordinarily, the carbon fibres used here come from fibre recyclate, for which there are hardly any applications as of yet.

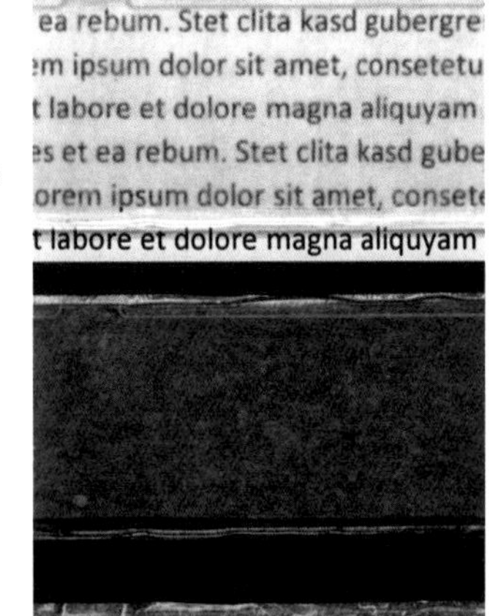

9 Paper-glass laminate with fluorescent fibres in the paper (above: illuminated by daylight against a white background with writing; centre: illuminated by daylight against a black background; below: illuminated by UV light against a black background).

The paper manufacturing processes described in **» chapters 2, 3** offer a wide range of fibre types for paper production. With them, fibre selection can be tailored for specific applications to influence the overall fibre properties. We know which fibres and treatment methods can be used to achieve high strength properties. The material data developed in this book were generated on the basis of this knowledge. However, building applications require additional parameters to be taken into account, such as thermal insulation and climatic properties. Generally speaking, porous structures insulate better than dense structures, and paper materials can be used to exploit this physical law. However, so far, there is little knowledge about how fibre pre-treatment and paper production affect the thermal insulation properties of papers – the potential for research and development is correspondingly unexplored. It is also known that paper fibres have a high capacity to absorb and release moisture. This factor could have a significant effect on the humidity levels of the indoor climate. So far, there are hardly any research findings on this either.

11 Sandwich core made of paper in Miura fold.

The use of filler materials, especially mineral ones, plays a major role in conventional paper manufacturing. The main aim is to improve visual properties and surface qualities; increase specific weight; and optimise resistance to water, grease, oil, etc. (barrier quality). Building applications can give additional reasons to improve the properties of paper, for example, in terms of thermal conductivity, fire protection, mass, sound insulation, and also resistance to moisture and microorganisms. The bonding ability of paper-based layers to mineral layers can also be improved by – for paper-concrete laminates, for example – the targeted introduction of fillers into the paper layer facing the (in this case) concrete surface.[1] Filler materials could, for example, be used to adjust a separation layer between a paper layer and a plaster surface in such a way that the plaster layer can be easily separated at the end of its useful life and recycled separately from the paper layer.

The field of material combinations – for example, in the form of laminates – is generally promising. Paper is very easy to combine with different materials by using the many hydroxyl groups on the surface of its cellulose fibres, which achieve good adhesive properties with a wide variety of adhesives or binders. Initial trials were conducted at TU Darmstadt to investigate possible paper-glass laminates »**fig. 8**. Glass and paper have complementary properties mechanically and on other physical levels, and thus offer the potential for exciting synergies. Both materials expand only slightly with increasing temperatures, and the visual properties of paper can be adjusted within relatively wide limits by employing suitable fibres and pre-treatment methods. It is thus conceivable to use paper as a reinforcing fibre in paper-glass laminates and to create diffuse light of different intensities by choosing paper of different translucency.

It is also relatively easy to embed functionalised fibres in the paper layer. In the example shown in »**fig. 9**, fluorescent fibres were incorporated into the paper layer. In daylight, the paper-glass laminate appears relatively transparent; in UV light, only the fluorescent fibres in the paper are visible. The bond between the paper and glass layers was successfully tested with both epoxy resin and polyvinyl butyral, a commonly used adhesive in the production of laminated safety glass.

Conventionally manufactured papers always have a preferred direction resulting from the production process, as most of the fibres in the paper machine are oriented in the direction of production »**chapter 2, pp. 28–29**. Experience with fibre composite materials has shown that a high level of unidirectionality of the reinforcing fibres in the direction of the load enables, in particular, very light constructions. Paper offers a lot of potential to fully exploit this basic principle of fibre orientation, as the orientation ratio of fibres in its longitudinal and transverse directions is in the order of 4:3. Therefore, a concept to produce paper webs with a high degree of fibre orientation was developed and tested at TU Darmstadt on a laboratory scale. In addition to the production of papers with high fibre orientation, this manufacturing principle also enables the incorporation of other fibres (for example, carbon fibres and fluorescent fibres »**fig. 10**).

Sandwich cores made of paper »**fig. 11** are also suitable for construction finishing and lightweight construction, especially furniture construction.[2] They enable very stiff and light constructions. Forming, embossing or folding techniques can be used to generate core structures which, on the one hand, can be rear-ventilated and, on the other hand, achieve similar strength characteristics in both surface dimensions. Classic corrugated board is one such sandwich core structure; however, mechanically, it is much weaker parallel to the flutes than perpendicular to them. Unfortunately, there are hardly any scientific studies or research projects on this type of paper.

Finally, it should be mentioned that conventionally produced multi-layer papers have neither been optimised nor systematically investigated for specific use in building construction. This opens up another large "playing field" for further innovations.

Recyclability

Paper in its pure form can be recycled without any problems. The recycling techniques required have existed for several centuries and have been constantly developed and optimised in order to make the fibres usable again at the best possible quality. However, material combinations and additional functions pose new challenges for future recycling processes of construction-optimised papers.

One typical example is fire protection – there are several strategies to optimise papers in this respect, most of which are based on material combinations. One variant is adding additives such as phosphorus-containing chemicals or mineral fillers to the pulp, i.e. the paper mass. This creates protection at the fibre level, and the carbon formation of the paper itself is promoted and stabilised. Alternatively, intumescent coatings that foam up in case of fire and protect the material behind them can be applied to the surface of the manufactured paper. Both additives and coatings make the recycling process more difficult because such additives, fillers or coatings must be separated from the paper fibres and filtered out of the process water.

Moisture protection, in particular, is an exciting and controversial topic concerning the durability of the building structure and its recyclability at the end of its life cycle. On the one hand, paper structures should be stable during their period of use and ensure a hygienic indoor climate. Moisture may not damage or even dissolve them during their useful lifetime. On the other hand, moisture in the form of water must be able to dissolve them during the recycling process and break them down into their fibres without any loss of material. This balancing act can also be solved with material combinations. An additional film-like layer on the surface of its paper, for example, can protect a construction from water penetration and be mechanically detached again at the end of the life cycle. This method makes moisture protection and recyclability equally possible. Another approach is to develop a moisture-protective coating or additive that chemically bonds with the paper fibres and dissolves with another chemical process at the end of the life cycle.

Ways will be found to combine several functions in one layer, for example, by using coatings or additives that make the paper water-repellent, fire-resistant and antimicrobial without limiting recyclability. In the near future, however, the solution will be to layer these functions – which, in turn, means adding additional process steps to the recycling process. These could include upstream mechanical processes that detach functionalised layers from the paper structure or additional chemical processes. One example is deinking, which is used to separate the ink from the fibres when recycling newsprint. In the future, there will be techniques that, for example, first remove or filter out the water-repellent and antimicrobial layer. The mixture could then be placed in a water bath where bacteria or similar degrade the fire retardants, after which the paper fibres could be recovered using conventional treatment processes.

In the future, however, not only functionalisation and recycling processes but also architectural planning must be adapted to temporary and circular constructions. The design process must not merely consider the end of the life cycle but bring it into focus.

Rethinking processes

The change in architecture and construction associated with new material lifetimes and durability requires a new thought process in the entire chain, from design and first idea to the realisation and use phase. Architects, engineers, manufacturers and processors alike must adapt to this change. It is not the longevity of a construction method that could be in the foreground of a design-planning task, but rather the exact opposite: processes and value chains that start from a relatively short circular process and work on a regenerative basis.

To achieve this, the focus must lie on the building materials to be used, and their raw materials must be closely examined. New value chains can be generated by, for example, recycling paper material or semi-finished products previously used as packaging material and using the recyclate to produce new semi-finished paper products for building construction. In this way, removing raw material mass from one value chain means creating another value chain.

This process can also be transferred to the raw material wood. Thus, the cascade use of wood (solid wood, chip-based products, fibre-based products, chemical products, energy recovery) at the level of fibre-based products can be expanded to include the use of wood fibres as a raw material for paper. Since paper can achieve similar properties to solid wood, this step is an upcycling of the material. The idea would be to generate wood-based materials that can be used several times before they are disposed of – usually thermally.

Perhaps it is precisely those properties of paper and cardboard that are still negatively perceived, e.g. short lifespan, that open up an undreamed-of potential for a different concept of architecture. Responding to a fluctuating market and to a growing housing demand with temporary units or one-to four-year paper structures in the form of lightweight constructions requires new thought and production processes, and a questioning of the status quo. Building with paper needs new planning processes; faster, systematised industrial processing; and, in transition, simple on-site construction processes. More prefabrication, less assembly and shorter construction times. More systematic thinking and working, less individual production. Or, in contrast, does the nature of the material and its processing make it particularly suited for individualised shapes and applications? Is a "plot" architecture, i.e. a construction method that grows out of machine production and processing, conceivable?

The material of the future must respond to all these aspects. Prefabrication and low weight offer potential for responding quickly to the demands and requests of a changing architectural market. New processes aim to use a material that consists entirely of renewable components and whose dismantling and disposal, or even reuse for new process chains, represents added value.

REFERENCES

1 Frederic Kreplin, Samuel Schabel, Martin Lehmann, Andreas Büter, Mandy Thomas, Tiemo Arndt, Sabrina Mehlhase, Markus Biesalski, Christian Mittelstedt, Andreas Maier, Nihat Kiziltoprak, Jens Schneider, Albrecht Gilka-Bötzow, Mona Nazari Sam, E. A. B. Koenders, Samuel Schabel (eds.), *KOMPAP – Energieeffizientes Bauen durch Komposit-Materialien mit Papier,* Darmstadt: TU Darmstadt, 2020, BMWi 03ET1414A-C + E [Report]

2 Niklas Schäfer, *Leichtbaupotential durch Faltstrukturen aus Papier als Sandwichkern*, Master's thesis, TU Darmstadt, 2020.

8 FACTS AND FIGURES FOR ENGINEERS

This chapter provides an overview of test procedures and test methods for paper materials, mechanical and biophysical component tests and fire protection. The information is intended as supplementary knowledge, especially for engineers, providing an in-depth understanding of the material and the tests necessary for building construction.

Test methods for papers and cardboards

A wide variety of test methods are used in the scientific context to determine the structural material parameters of different paper materials. Not all of these methods are standardised. The focus of this chapter lies on test methods that are established in the industry and do not require special equipment. A very comprehensive overview of mechanical material properties is provided in the *Handbook of Physical Testing of Paper.*[1]

Sampling, pre-treatment and base values

Sampling can influence the result of a test, which is why sampling is standardised in DIN EN ISO 186.[2] However, samples cannot always be chosen under one's own auspices to comply with the standards but are supplied by a third party. In this case, the measurement results must be interpreted accordingly. Before further testing, the samples must be pre-treated in standard climate conditions for up to 48 hours for the sample to reach the standard temperature and moisture content.

As described in the »**section "Paper, paperboard and cardboard", pp. 36–38**, grammage is one of the most important parameters for characterising paper. It is determined by weighing a sample of a defined size (see DIN EN ISO 536).[3]

Calculating tensile strength, for example, requires information on the paper's thickness. Due to the small number of fibre cross-sections lying on top of each other in a sheet of paper, determining the thickness is critical. According to DIN EN ISO 534, the thickness can be determined on a single sheet or on a stack of sheets using a micrometer that applies pressure to a defined area.[4]

Tensile testing

In tensile tests, a distinction is made between tests in the direction of paper web orientation during production (i.e. machine or in-plane direction), on the one hand, and in the cross direction (out of plane), on the other. Tensile testing of paper, in general, is described in DIN EN ISO 1924-2.[5]

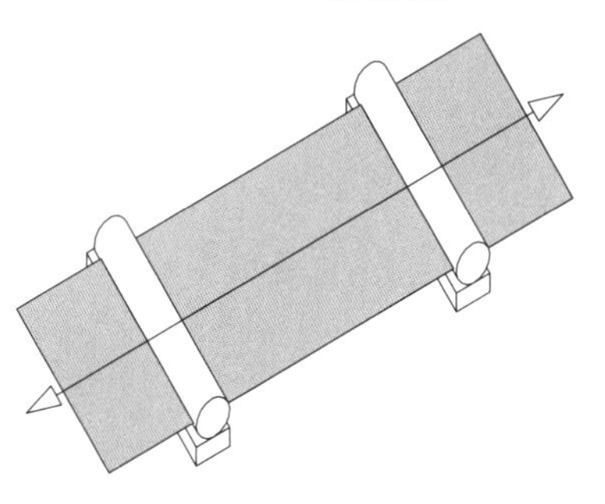

1 Schematic diagram of tensile testing in the machine (in-plane) direction.

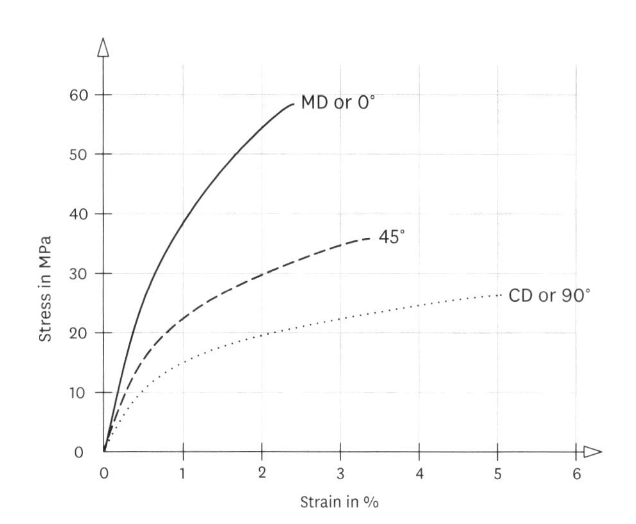

2 Typical stress-strain curve for paper samples 0°, 45°, 90°.

In-plane testing

In contrast to other materials, in-plane tests for paper use rectangular strips »**fig. 1** that are 15mm wide and – unlike other test specimens – have straight edges. »**Fig. 2** depicts a typical stress-strain curve.

The curve shows a small, purely elastic area; followed by an elastic-plastic to a purely plastic area; and, finally, the tear. Not shown here is the visco-elastic material behaviour. If an initial strain is maintained, the stresses will reduce over time. In a cyclic tensile test, elongation regresses after the relaxation phase until the next load is applied. The modulus of elasticity is determined by applying a tangent in the elastic range or the range of the maximum increase of the force-elongation curve »**fig. 3**.

Other characteristic parameters are tensile strength and elongation at break, which can be read at the highest point of the curve. Traditionally, paper is often characterised by its breaking length. This means the minimum length a freely hanging strip of paper must have to tear under its own weight. This parameter simplifies the comparison of different papers. In the construction sector, the focus is on the elastic range of the tensile curve. It is difficult to give generally valid values or value ranges for paper because the purely elastic range cannot be identified as clearly as with other materials. It should be noted that paper develops higher strengths under dynamic loading, i.e. at high strain rates, than under slow loading.

The Poisson's ratio can be determined with an optical measuring device, a so-called extensometer. »**Fig. 4** shows the measurement result: the transverse strain is plotted against the longitudinal strain. The Poisson's ratio results from the gradient.

Paper shows an auxetic material behaviour »**fig. 5**. During the tensile test, it becomes thinner in the plane of the sheet while at the same time it expands in the through thickness direction. This effect results from the fibre network typical of the

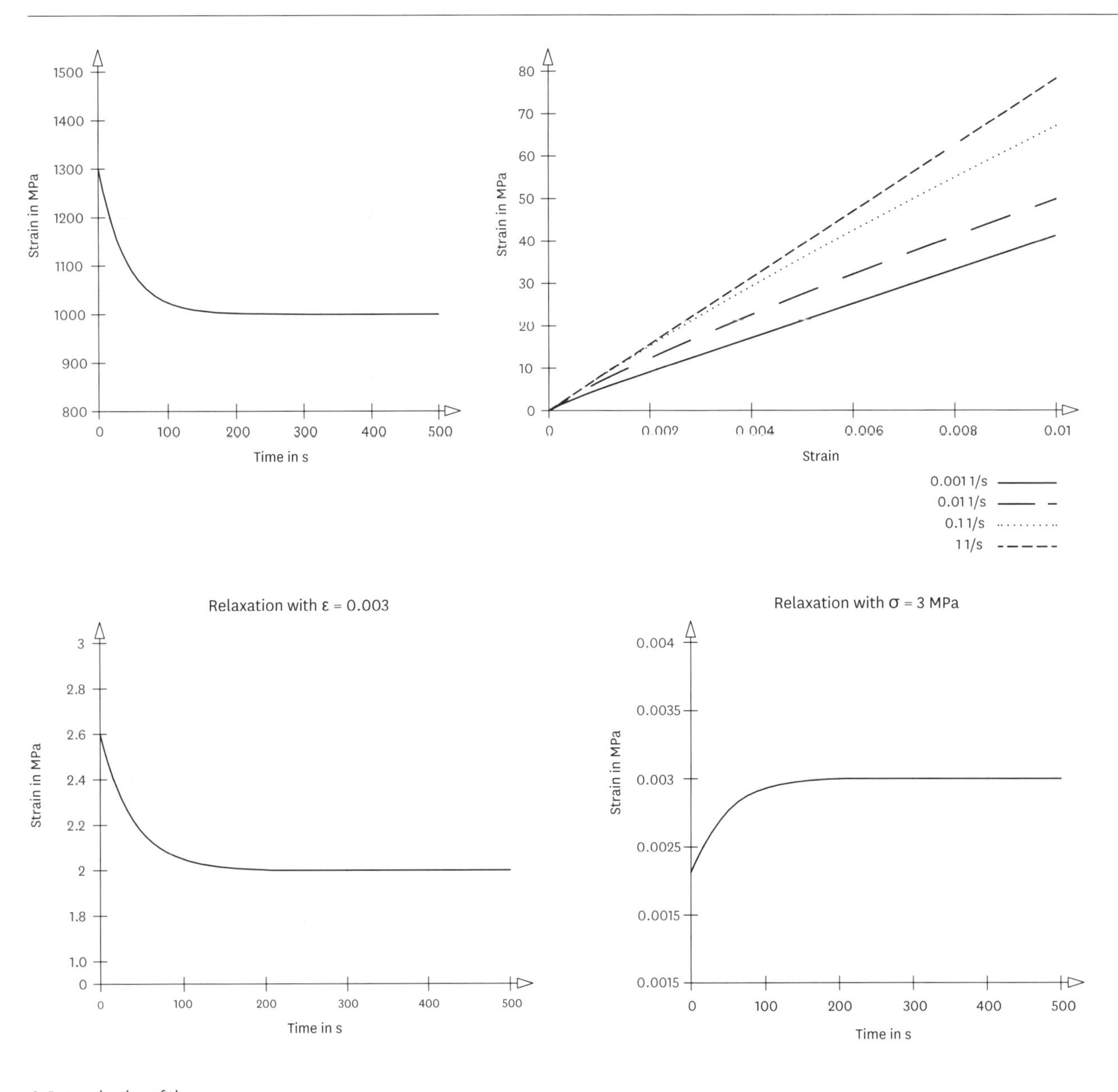

3 Determination of the modulus of elasticity, taking into account the influence of time as well as relaxation and creep.

material: the fibres that lie in the direction of the load are stretched and raise crossing fibres. When calculating the effective stress, a change in the cross-section must be taken into account but the bearable load remains the same.

Due to the fibre orientation in the paper, the test is carried out in machine direction MD (0°) and in cross direction CD (90°). The angles in between allow statements about the behaviour under compression.

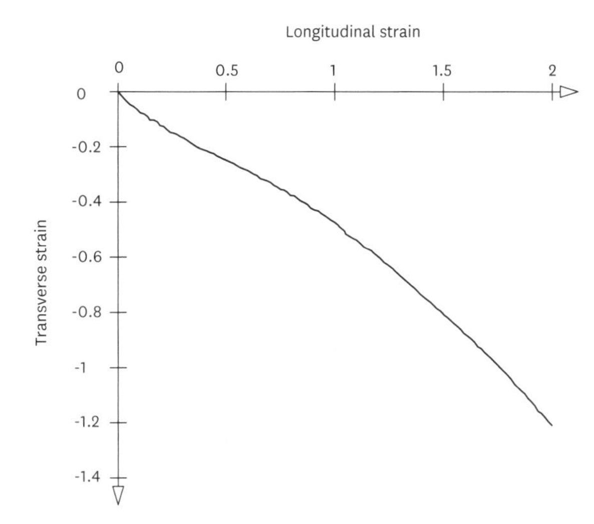

4 Transverse strain over longitudinal strain to determine the Poisson's ratio.

Test in cross direction

There are several standardised methods to characterise the splitting behaviour in classic paper applications. However, they follow two basic principles. In the TAPPI T 541[6] method, double-sided adhesive tape is applied to the top and bottom of the paper. The sample is then placed between two plates and pressed against them. The resulting adhesion is the prerequisite for subsequent uniaxial stretching in the testing machine.

The plybond resistance test, as stated in DIN 54516, requires a special sample holder to initiate the sample splitting from an edge.[7] In the Tappi T 569[8] method, splitting also starts from one edge, but in this second method a pendulum strikes an angle bar taped to the paper. The results of both methods are not directly comparable. Based on the first principle, a special sample holder can be set at different angles, which allows the sample to slide downward. The uniaxial tension is well suited to the requirements of construction applications as the conditions here are quasi-static, whereby the pendulum impact corresponds to a dynamic load. It is difficult to determine the Poisson's ratio with this test. In certain cases, however, it is permissible to assume the Poisson's ratio to be 0.[9]

Compression tests

Since paper is very thin, compression-testing it is complicated. With the in-plane method, the test specimens buckle easily; with an out-of-plane test, the thickness of the small test specimens requires very high measurement accuracy for the single-sheet test. An alternative is stack testing.

In-plane testing (MD)

DIN 54518 standardises the determination of the crushing resistance using the so-called Short-Span Compression Test (SCT).[10] Here, clamps on either end fasten a sample [illegible] The distance

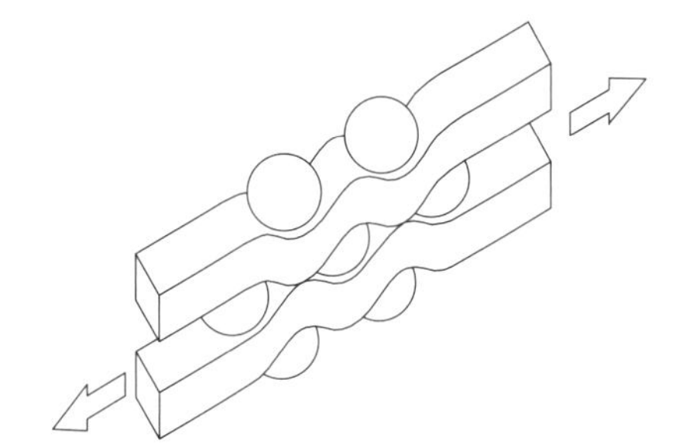

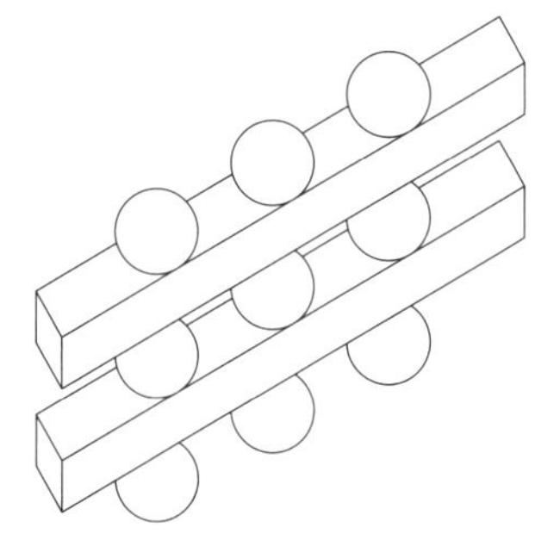

5 Auxetic material behaviour of the fibre network of paper: when tensile loads stretch the crossing fibres, the paper tapers in the sheet plane and expands in thickness.

between the clamps is only 0.7mm, intended to prevent the paper from bulging. The crushing resistance is the maximum value of the force occurring when the clamps are moved together, related to the sample width. This test also distinguishes between MD and CD. The thinner the paper, the more likely it is to buckle.

Other methods are used for corrugated base paper testing. Here, the test strip is placed in a ring (Ring Crush Test, RCT) or pressed into a corrugated sample holder (Corrugated Crush Test, CCT). The curvature of the specimens prevents buckling but influences the measured values. Results of these tests showed that the modulus of elasticity in the elastic range under compression is equal to the modulus of elasticity under tension.[11]

Out-of-plane testing (CD)

There are no standardised compression tests for testing paper perpendicular to the sheet plane. One possible test set-up is that used by Jian Chen, which consists of a universal testing machine with a load sensor, a base plate, a compression die with a spherical cap, and an extensometer.[12] The test specimen and the extensometer are located between the base plate and the compression die **»figs. 6, 8**. The spherical cap guarantees parallel alignment of the surfaces. Hereby, it is often assumed that – as with out-of-plane loads – the Poisson's ratio is negligible.

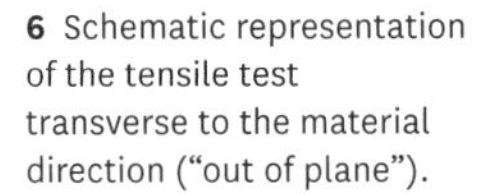

6 Schematic representation of the tensile test transverse to the material direction ("out of plane").

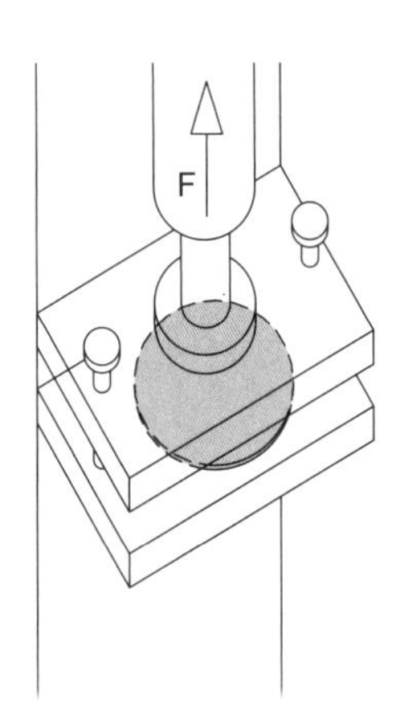

Test methods for mechanical failure of components

To determine the suitability of components as load-bearing elements, corresponding load situations must be tested. If the load-bearing capacity of a component cannot be calculated, experimental test methods need to be conducted. The type of test depends on the parameter sought or on the type of expected load on the component.

Tensile test

DIN EN 408, which relates to timber structures, lends itself to standardising tensile tests at the component level.[13] A corresponding test device must accommodate a test specimen length that is at least nine times larger than the cross sectional dimension.

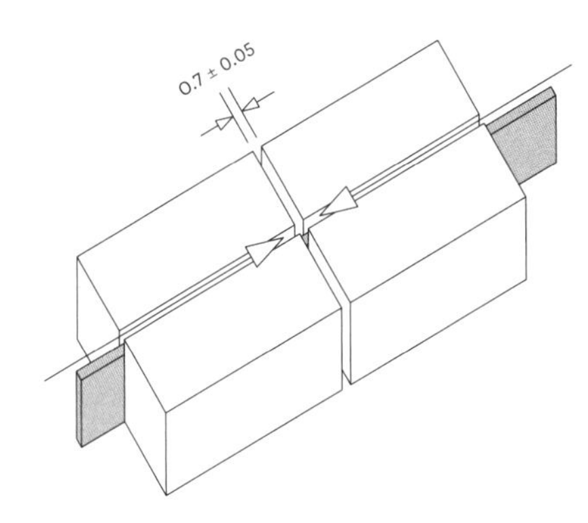

7 Schematic representation of the compression test in the material direction ("in-plane").

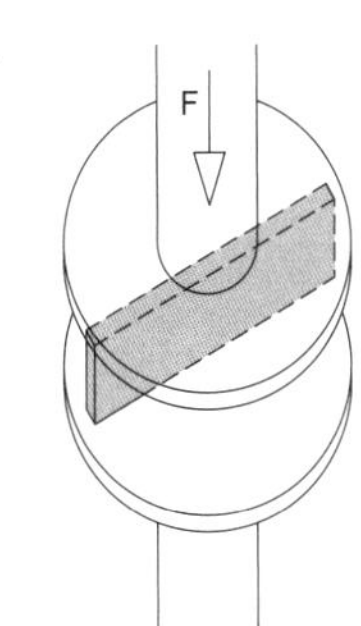

8 Schematic representation of a potential compression test transverse to the material direction ("out-of-plane").

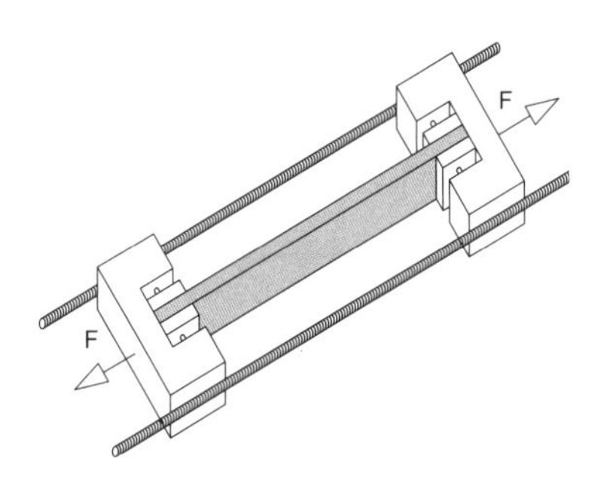

9 Schematic representation of a component subjected to tensile testing.

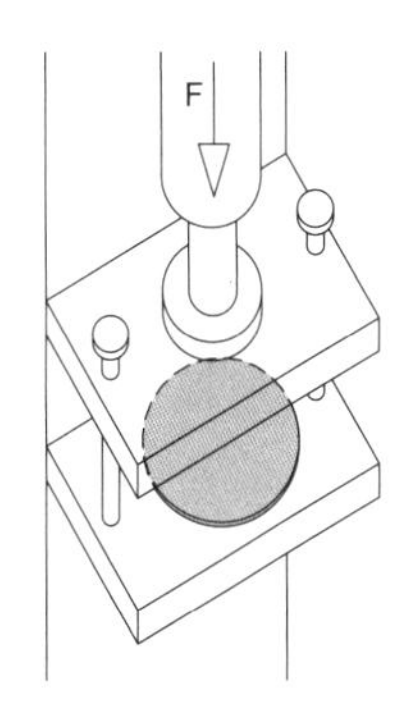

10 Schematic representation of a component subjected to compression testing.

In addition, the test specimen must be clamped in such a way that it cannot bend during the test »**fig. 9**. For wood, the speed of a strain-controlled test is specified at 0.005%/s. If this methodology is transferred to paper components, the parameters may have to be adjusted. The tensile test can be used to determine the tensile strength and the tensile modulus of elasticity.

Compression test

In addition to provisions for tensile tests on components, DIN EN 408 also contains provisions for compression tests. The standard specifies, for example, that the test specimen length must be six times larger than the smaller cross-sectional dimension. The compressive force must act centrally on the test specimen without allowing it to bend. Spherical caps are ideal for this purpose »**fig. 10**. To prevent the test specimen from buckling in the bearing area, it must be appropriately secured after the load has been applied. This method can be used to determine the compressive modulus of elasticity and the compressive strength. It also enables the component to be designed as slender as possible without compromising stability.

Bending test

Bending tests are carried out on both beam-like and flat components, such as ceiling slabs. The components usually rest on supports at both ends and are stressed in the area between them **»fig. 11**. Both the load and the deformation that occur during the load are measured. Variations such as three-point, four-point and six-point bending are common. Four-point bending is suitable for determining an almost constant moment distribution between the two loading points. Six-point bending is often used to simulate load distribution across the span of the component.

In the building industry, DIN EN 384, for example, refers to bending tests of components in timber construction.[14] But standards also already exist for bending tests on paper tubes. DIN ISO 11093-6[15] and DIN ISO 11093-7[16] give recommendations both for the load and bearing distances to be observed and for the geometry of the load application elements (bearings and force points) to avoid so-called ovalisation of the cross-section during loading. Local effects can be eliminated by using the largest possible contact surfaces on the load application elements. It should also be noted that, during testing, the load application elements can press into the test specimen, which makes it necessary to conduct deformation measurements using external displacement sensors.

11 Schematic representation of a component subjected to bending testing.

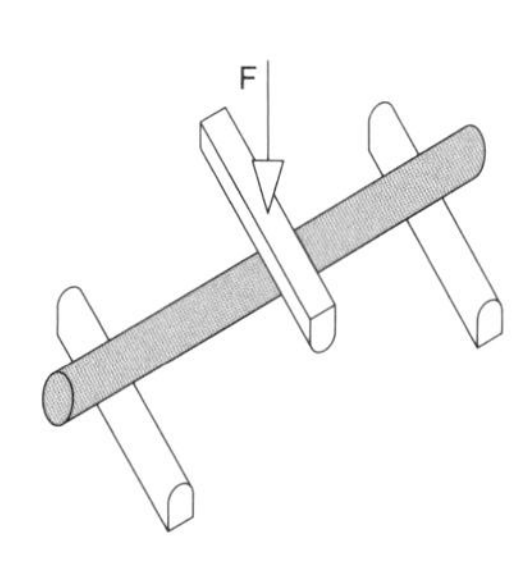

12 Schematic representation of a component subjected to torsion testing.

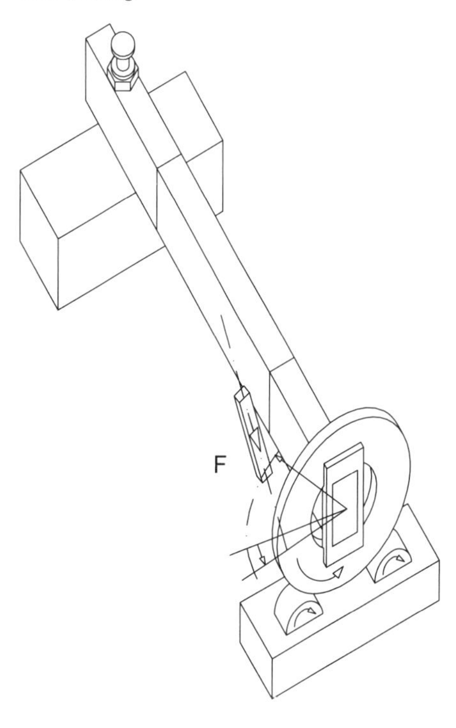

13 Schematic representation of a component subjected to shear testing with a tension-compression testing machine.

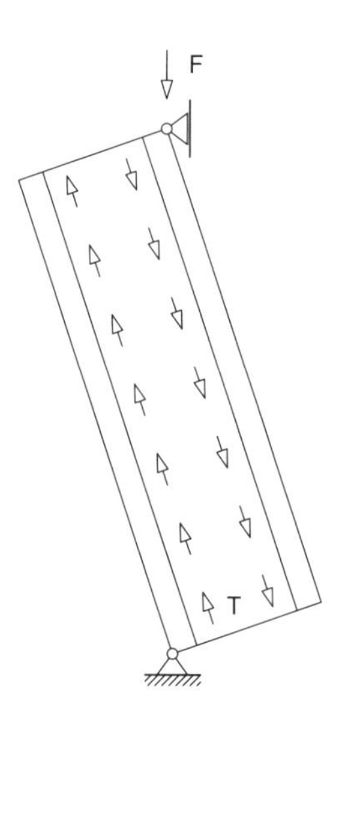

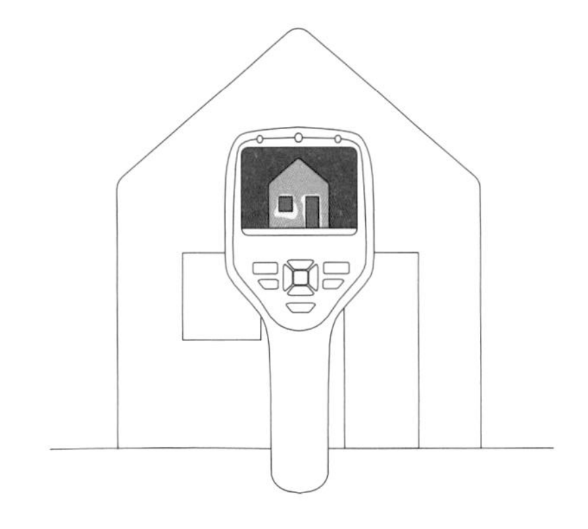

14 Thermal bridges can be identified with thermal imaging cameras.

In addition to the bending stiffness of the component, bending tests can be employed to determine other strength and stiffness values, such as the bending tensile strength, bending compressive strength and the modulus of elasticity in the longitudinal direction of the component. Depending on the material and component dimensions, shear deformations may occur, which may result in an underestimation of the bending stiffness. The influence of the shear force can be estimated by bending the beams at different bearing and load distances.

Torsion test

DIN EN 408 also describes a possible device for torsion testing on components. Accordingly, the component is clamped at one end in such a way that rotation about the longitudinal axis of the test specimen is prevented. However, the bearing at the other end allows relative rotation »**fig. 12**. For this purpose, one end is clamped in a cylinder, for example, which rests on two other cylinders and can be rotated about the longitudinal axis of the test specimen by means of a lever arm.

For timber components, the test specimen length is specified as 19 times the component height. For paper, a suitable value is to be investigated. The torsion test can be used to determine both the torsional moment that can be absorbed by the component and the shear modulus.

Shear test

The shear properties of beam-type components can be determined with torsion tests. Alternatively, shear tests can also be carried out experimentally with the aid of a ten-sion compression testing machine. However, this requires appropriate attachments

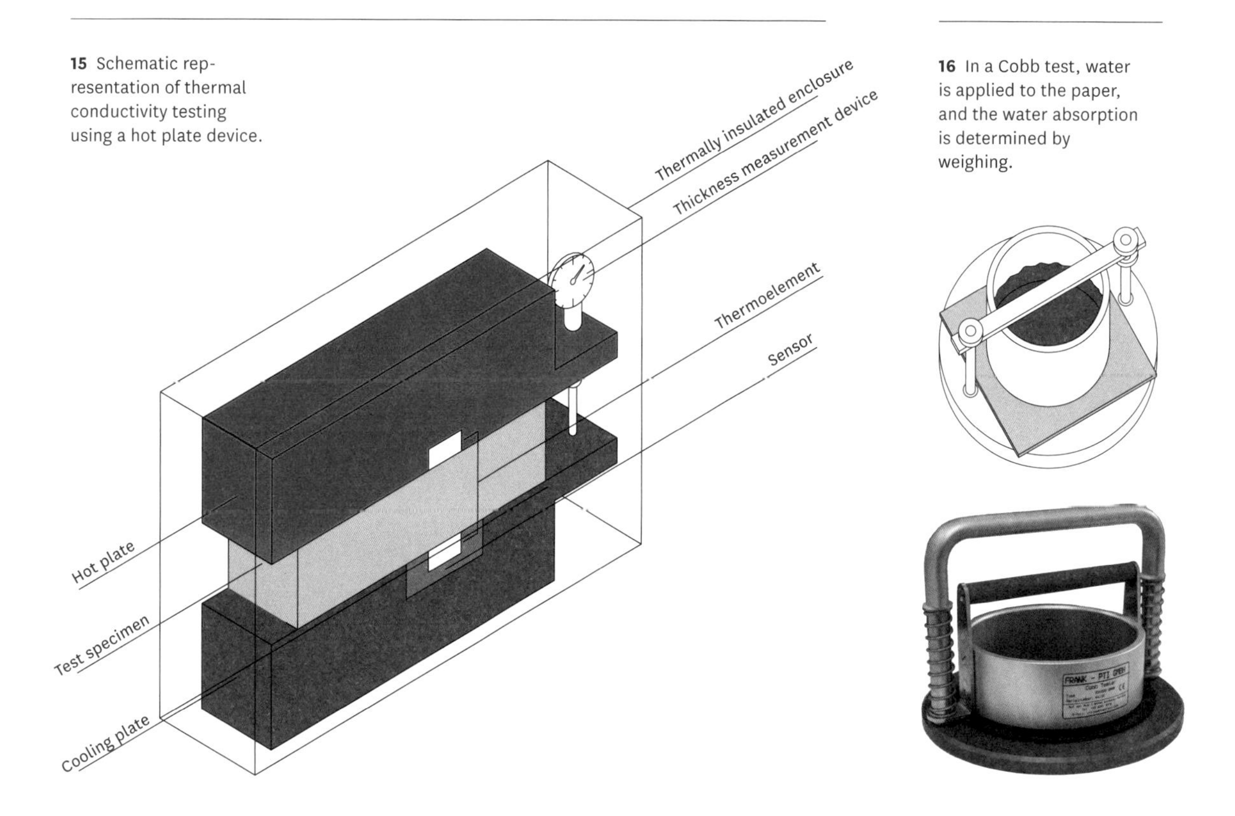

15 Schematic representation of thermal conductivity testing using a hot plate device.

16 In a Cobb test, water is applied to the paper, and the water absorption is determined by weighing.

for the testing machine: sufficiently stiff plates, which are hinged to the two connection points of the testing machine, are mounted on the top and bottom of the test specimen » **fig. 13**. Shear force is then applied by the vertical travel of the testing machine. Corresponding physical values are to be determined via mechanical relationships.

Test methods for building physics

In addition to the aforementioned test methods for determining the mechanical material properties of building components and their load-bearing capacity, building physics tests are essential as well. The thermal conductivity of the paper materials in building envelopes ultimately determines the level of comfort and the energy balance of the building. To prevent the construction from being damaged by water and mould, the surfaces of the building components must be protected accordingly. In this context, the water vapour diffusion density of the component layers is also relevant – as is fire protection.

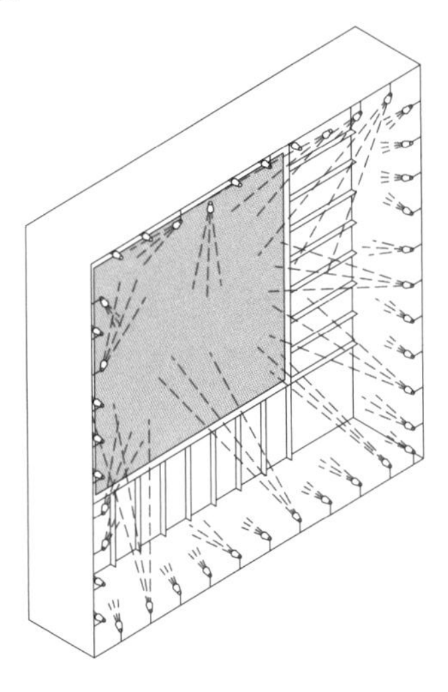

17 Driving rain tests on the façade test stand can be used to verify a construction's watertightness.

Thermal conductivity determination

The thermal conductivity of materials can be determined using stationary or dynamic methods. For paper materials, the most suitable stationary methods are those which determine the thermal conductivity with guarded hot plates »**fig. 15**, heat flow meters or by means of the relative method. These tests are carried out according to DIN EN 12667.[17] Tempering one or more plates generates heat flow in the material, which can then be measured. The thermal conductivity of the material can be derived from this heat flow. Thermal imaging cameras are suitable for determining thermal bridges and surface temperatures »**fig. 14**.

Cobb test to determine the water absorption

According to DIN EN ISO 535:2014-06, the so-called Cobb test »**fig. 16** can be used to determine water absorption capacity – i.e. the absorbency of paper materials.[18] For this purpose, water is applied to the paper over a defined period of time. The paper is

18 Schematic representation of water vapour diffusion tests using the cup method.

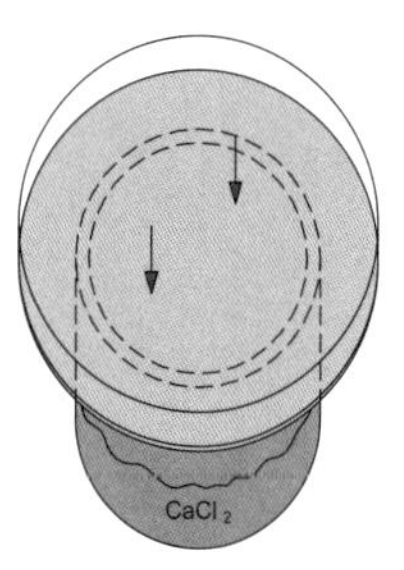

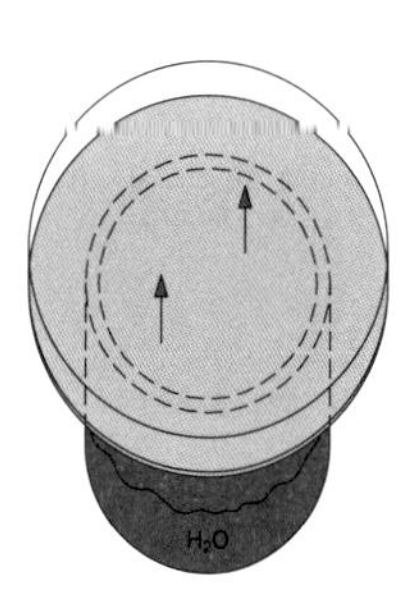

19 Schematic representation of single-flame source tests.

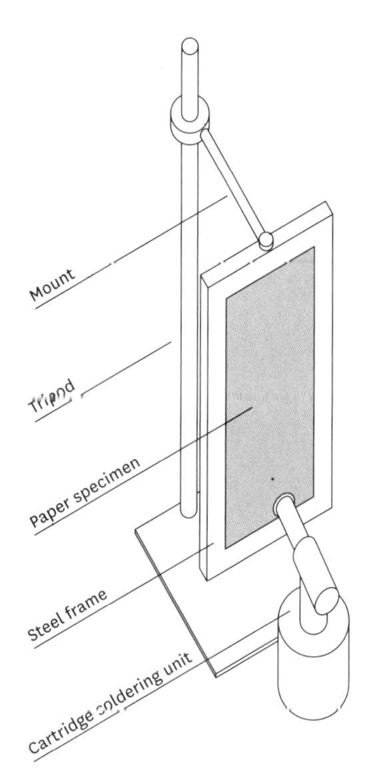

then weighed, which gives the water absorption in g/m². The watertightness of entire constructions can be verified in façade test stands by means of driving rain tests conducted according to DIN EN 12865.[19] In addition to a constant sprinkling of the test specimen, pulsating air pressure is used to simulate the impact of water on the construction at different angles » **fig. 17**.

Water vapour diffusion tests

Not only the water absorption capacity of paper but also its water vapour diffusion density plays an important role in construction. The water vapour diffusion coefficient, which is given in μ independently of thickness, is determined according to the standards DIN EN ISO 7783[20] and DIN 53122-1[21] with so-called wet and dry cup testing » **fig. 18**. For this purpose, cups are filled with dry and wet media and sealed with the paper material to be tested. Desiccants create a very low humidity inside the cup so that ambient humidity diffuses through the paper into the cup to create an equilibrium. A wet medium, on the other hand, creates high humidity so that the water vapour diffuses through the paper out of the cup. The cups are weighed regularly over a known period, allowing the water vapour diffusion flow to be determined. This, in turn, results in the thickness-independent permeability of the material.

Fire protection tests

To classify paper materials in the fire protection classes according to DIN EN 13501[22] and DIN 4102,[23] single-flame source tests »**fig. 19** are carried out according to DIN EN ISO 11925.[24] In the course of these tests, the materials are flame-treated for a defined period of time. The classification is determined by flammability, burning rate, smoke development, burning drippings and other characteristics. To determine the fire resistance class of building components, these are tested by a fire protection laboratory according to their installation situation in accordance with the DIN EN 1363–1366[25] standards.

Material samples

The materials represent a selection of common paper materials:

- Combination of two types of paper: liner and fluting paper. These two paper grades are the basic materials for the semi-finished products corrugated board and corrugated multi-wall board presented in »**chapter 3, pp. 38–41**. Depending on the axis direction (MD = machine direction, CD = cross direction), the values are different and, therefore, both ratios are given, plus the value for the diagonal between MD and CD (45°) for obliquely acting loads »**figs. 20, 21**.
- Paperboard. The selected paperboard is a three-ply board consisting of a white outer ply of bleached fresh fibre, a grey inner ply of mixed waste paper and a middle ply of unbleached waste paper containing wood pulp. This paperboard is very pliable »**fig. 22**.
- Cardboard. Cardboard is a material comparable with paperboard »**fig. 22**, but with a higher grammage »**fig. 23**. In this case, however, the cardboard does not consist of different types of paper that have been couched together, like the paperboard in »**fig. 22**, but only of wood pulp.
- Corrugated board. Here, liner and fluting paper are combined, comparable with the materials in »**fig. 20** and »**fig. 21**. The special feature lies in the structure of the material, which has different mechanical properties depending on its orientation »**fig. 24**. The corrugated structure between the liners enables high stability at a low weight.
- Corrugated multi-wall board. For corrugated multi-wall board, fluting paper is cut open and lined up vertically. The source material is therefore identical to the material in »**fig. 24**; only the structural orientation varies and causes the differences in the mechanical properties »**fig. 25**.

Material example Liner

20 Liner. The material is used in combination with corrugated paper (fluting) » **fig. 21**.

Grammage		135g/m²
Thickness		0.18mm
Density		754kg/m³
Tensile stress	MD	67.69 MPa
	45°	40.14 MPa
	CD	29.01 MPa
Tensile strength	MD	181.24 N
	45°	107.47 N
	CD	77.67 N
Tensile index	MD	1.35Nm²/g
	CD	0.58Nm²/g
Breaking length	MD	9152.56m
	CD	3922.53m
Elongation at break	MD	1.71%
	45°	2.73%
	CD	5.45%
Modulus of elasticity	MD	6761.80 MPa
	45°	3907.6
	CD	2451
Poisson's ratio	MD	0.352
	45°	0.159
	CD	0.144
Shear modulus		1748.64 MPa
SCT strength	MD	4.48 kN/m
	CD	2.80 kN/m

Material example Fluting

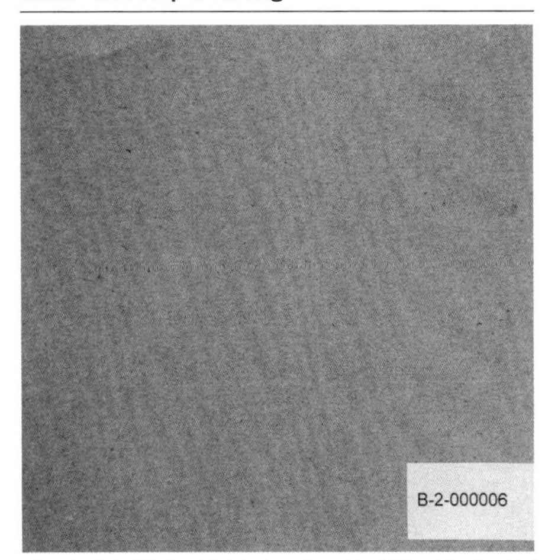

21 Corrugated paper (fluting). The material is combined with a cover paper (liner) » **fig. 20**.

Grammage		139g/m²
Thickness		0.20mm
Density		687kg/m³
Tensile stress	MD	30.86 MPa
	45°	19.06 MPa
	CD	14.34 MPa
Tensile strength	MD	93.74 N
	45°	57.89 N
	CD	43.56 N
Tensile index	MD	0.67 Nm²/g
	CD	0.31 Nm²/g
Breaking length	MD	4579.68 m
	CD	2128.08 m
Elongation at break	MD	1.34%
	45°	2.01%
	CD	3.18%
Modulus of elasticity	MD	4061 MPa
	45°	2660.1
	CD	1914
Poisson's ratio	MD	0.342
	45°	0.229
	CD	0.191
Shear modulus		1106.9 MPa
SCT strength	MD	3.83 kN/m
	CD	2.54 kN/m

Material example Paperboard

22 Three-ply paperboard. Paperboard differs from paper in its higher grammage. Compared with cardboard, on the other hand, the grammage is lower. Paperboard is usually made by couching several paper layers. Combining different types of paper (here, a white outer ply of bleached fresh fibre, a grey inner ply of mixed waste paper and a middle ply of unbleached waste paper containing wood pulp) can produce different paperboard qualities.

Property	Direction	Value
Grammage		385g/m²
Thickness		0.53mm
Density		728kg/m³
Tensile stress	MD	59.40 MPa
	45°	34.94 MPa
	CD	25.41 MPa
Tensile strength	MD	471.87 N
	45°	277.56 N
	CD	201.86 N
Tensile index	MD	1.22 Nm²/g
	CD	0.52 Nm²/g
Breaking length	MD	8318.96 m
	CD	3558.67 m
Elongation at break	MD	2.09%
	45°	3.03%
	CD	4.42%
Modulus of elasticity	MD	6114.08 MPa
	45°	3511.7
	CD	2213.34
Poisson's ratio	MD	0.466
	45°	0.169
	CD	0.202
Shear modulus		1479.02 MPa
SCT strength	MD	9.14 kN/m
	CD	5.98 kN/m

Material example Cardboard

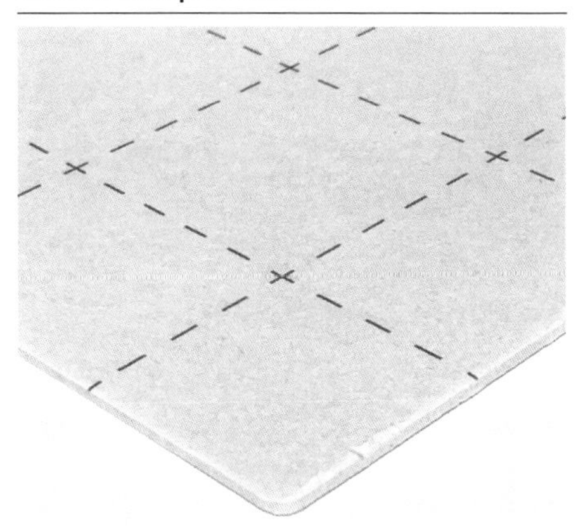

23 Cardboard. Cardboard has the highest weight per unit area, or grammage, compared with paper and paperboard. Cardboard is usually made by couching several papers. Its high volume makes it suitable for building construction applications. Compared to paper, significantly fewer layers need to be glued together. The product shown here is already used as impact sound insulation. It is moisture-resistant due to wet-strength agents.

Grammage		708g/m²
Thickness		1.81mm
Density		391kg/m³
Tensile stress	MD	6.02 MPa
	45°	4.72 MPa
	CD	4.16
Tensile strength	MD	163.66 N
	45°	128.32 N
	CD	113.09 N
Tensile index	MD	0.23 Nm²/g
	CD	0.16 Nm²/g
Breaking length	MD	1570.21 m
	CD	1085.06 m
Elongation at break	MD	0.86%
	45°	1.09%
	CD	1.52%
Modulus of elasticity	MD	950.92 MPa
	45°	705.11
	CD	534.17
Poisson's ratio	MD	0.095*
	45°	0.065*
	CD	0.095*
Shear modulus		340 MPa*
Bending stiffness (2-point method)	MD	516.56 Nmm
	CD	261.72 Nmm

* Ambiguities in measurement – values only indicative

Material example Corrugated board

24 Corrugated board. Corrugated board is made from the materials liner »**fig. 20** and fluting »**fig. 21**. Liner and fluting can be combined in several layers to form single to triple wall corrugated boards. The material properties of corrugated board vary depending on the design of the respective liner or fluting paper.

Cover paper (liner)	»**fig. 20**
Fluting paper	»**fig. 21**
Flute type	C-flute
Mass per unit	474g/m²
Thickness	3mm
Edge crush resistance	5.18 kN/m
Edge crush resistance material rotated by 90°	1.40 kN/m
Flat crush resistance	226.7 kN/m²

Material example Corrugated multi-wall board

25 Corrugated multi-wall board. The core of corrugated multi-wall board is made from corrugated board that is cut to size and glued together. Then, these so-called corrugated cores are covered with liner papers as in »**fig. 20**.

Mass per unit area	3450g/m²
Thickness	30mm
Density	115kg/m³
Compression test perpendicular to the plane of the cover layer	0.8 MPa

REFERENCES

1 R. E. Mark, C. C. J. Habeger, J. Borch, M. B. Lyne (eds.), *Handbook of Physical Testing of Paper,* vol. 1, New York, Basel: Marcel Dekker, Inc. 2002.
2 *DIN EN ISO 186:2002-08. Paper and Board – Sampling to Determine Average Quality,* Berlin: Beuth, 2002.
3 *DIN EN ISO 536:2019. Paper and Board – Determination of Grammage,* Berlin: Beuth, 2019.
4 *DIN EN ISO 534:2011. Paper and Board – Determination of Thickness, Density and Specific Volume,* Berlin: Beuth, 2011.
5 *DIN EN ISO 1924-2:2008. Paper and Board – Determination of Tensile Properties – Part 2: Constant Rate of Elongation Method (20mm/min),* Berlin: Beuth, 2008.
6 Zwick/Roell, https://www.zwickroell.com/en/branches/paper-card-board-tissues/paper/ internal-bond-test-tappi-t-541-iso-15754/, accessed 10 May 2022. See also Ahmed Koubaa, "Measure of the Internal Bond Strength of Paper/Board", *Tappi Journal* 78(3), March 1995, pp. 103–111. https://www.researchgate.net/publication/233898670_Measure of the internal bond strength_of_paperboard, accessed 6 [illegible]
7 *DIN 54516:2004-10. Testing of Paper and Board – Determination of Plybond Resistance,* Berlin: Beuth, 2004.
8 Tappi, "Internal Bond Strength (Scott Type), Test Method TAPPI/ANSI T 569 om-14". https://imisrise.tappi.org/TAPPI/Products/01/T/0104T569.aspx, accessed 10 May 2022.
9 N. Stenberg, C. Fellers, "Out-of-plane Poisson's Ratios of Paper and Paperboard", *Nordic Pulp & Paper Research Journal,* 17(4), 2018, pp. 387–394. doi:10.3183/npprj-2002-17-04-p387-394, accessed 4 October 2019.
10 *DIN 54518:2022-01. Testing of Paper and Board – Compression Strength, Short Span Test,* Berlin: Beuth, 2022.
11 K. Niskanen, "Structural Mechanics of Paper and Board", in: I. Kajanto (ed.), *Paper Physics,* vol. 16, Helsinki: Fapet Oy, 1998, pp. 192–221.
12 Jian Chen, *Investigation on the Mechanical Behavior of Paper and Paper Stacks in the Out-of-Plane Direction,* Dissertation, TU Darmstadt, 2016.
13 *DIN EN 408:2012-10. Timber Structures – Structural Timber and Glued Laminated Timber – Determination of Some Physical and Mechanical Properties,* Berlin: Beuth, 2012.
14 *DIN EN 384:2022-08. Structural Timber – Determination of Characteristic Values of Mechanical Properties and Density,* Berlin: Beuth, 2022.
15 *DIN ISO 11093-6:2005-09. Paper and Paperboard – Testing of Cores – Part 6: Determination of the Bending Strength by the Three-Point Method,* Berlin: Beuth, 2005.
16 *DIN ISO 11093-7:2012-10. Paper and Board – Testing of Cores – Part 7: Determination of Flexural Modulus by the Three-Point Method,* Berlin: Beuth, 2012.
17 *DIN EN 12667:2001-05. Thermal Performance of Building Materials and Products – Determination of Thermal Resistance by Means of Guarded Hot Plate and Heat Flow Meter Methods – Products of High and Medium Thermal Resistance,* Berlin: Beuth, 2001.
18 *DIN EN ISO 535:2014-06. Paper and Board – Determination of Water Absorptiveness – Cobb Method,* Berlin: Beuth, 2014.
19 *DIN EN 12865:2001-07. Hygrothermal Performance of Building Components and Building Elements – Determination of the Resistance of External Wall Systems to Driving Rain Under Pulsating Air Pressure,* Berlin: Beuth, 2001.
20 *DIN EN ISO 7783:2019-02. Paints and Varnishes – Determination of Water-Vapour Transmission Properties – Cup Method,* Berlin: Beuth, 2019.
21 *DIN 53122-1:2001-08. Testing of Plastics and Elastomer Films, Paper, Board and Other Sheet Materials – Determination of Water Vapour Transmission – Part 1: Gravimetric Method,* Berlin: Beuth, 2001.
22 *DIN EN 13501-1:2019-05. Fire Classification Assessment of Construction Products and Building Elements – Part 1: Classification Using Data From Reaction to Fire Tests,* Berlin: Beuth, 2019.
23 *DIN 4102-1:1998-05. Fire Behaviour of Building Materials and Building Components – Part 1: Building Materials; Concepts, Requirements and Tests,* Berlin: Beuth, 1998.
24 *DIN EN ISO 11925-2:2020-07. Reaction to Fire Tests – Ignitability of Products Subjected to Direct Impingement of Fire – Part 2: Single-Flame Source Test,* Berlin: Beuth, 2020.
25 *DIN EN 1366-1:2020-11. Fire Resistance Tests for Service Installations – Part 1: Ventilation Ducts,* Berlin: Beuth, 2020.

EDITORS AND AUTHORS

Professor Dr.-Ing. Ulrich Knaack studied architecture at RWTH Aachen University and, after several years in building practice and as a façade planner, is now Professor of Design of Construction at Delft University of Technology and Professor of Façade Structures at the Technical University of Darmstadt. Author of the well-known textbooks *Konstruktiver Glasbau* and *Konstruktiver Glasbau 2*, co-author of the "Imagine" and "REAL" book series and co-editor of the *Journal of Facade Design & Engineering*.

At Birkhäuser, he conceptualised the successful series "Principles of Construction", with four volumes published to date. He regularly organises conferences and exhibitions, such as PowerSkin, Future Envelope, glass technology live – the Hub, BE-AM (additive manufacturing in construction) and BAMP! – Building with Paper, a conference in connection with the BAMP! research programme funded by the State of Hesse (LOEWE).

Dr.-Ing. Rebecca Bach studied architecture at the Ostwestfalen-Lippe University of Applied Sciences in the Bachelor's and Master's programmes. She then worked as a research assistant on Europe-wide research projects at the Junior Professorship for Recyclable Construction at RWTH Aachen University. She received a doctoral scholarship from the German Federal Foundation for the Environment (Deutsche Bundesstiftung Umwelt) to investigate the use of paper materials in exterior wall constructions from a building physics and ecological perspective. Rebecca Bach became part of the BAMP! team and worked at the Institute of Structural Mechanics and Design at TU Darmstadt, first as a research assistant and then as a junior research group leader for "Energy Efficient Construction" and "Paper Construction and Design". In 2021, she completed her doctorate on the topic of paper façades. She has been working as a project manager for building physics and sustainability at Kempen Krause Ingenieure since 2020.

Professor Dr.-Ing. Samuel Schabel studied process engineering at Clausthal University of Technology, completed his doctorate at TU Kaiserslautern and then worked in the paper technology research and development department at Voith Sulzer. Samuel Schabel has been Professor of Paper Technology in the Department of Mechanical Engineering at TU Darmstadt since 2002. His work focuses on the processing of waste paper, the recycling of paper products and environmental issues relating to paper.

Chapter authors

1 Paper in Architecture
Ulrich Knaack, Jerzy Łątka, Fabian Luttropp, Marco Volkmann

2 Material
Cynthia Cordt, Oliver Elle, Andreas Geißler, Robert Götzinger, Felix Schäfer

3 Semi-Finished Products and Components
Robert Götzinger, Julian Mushövel, Sandra Schmidt, Paul Töws

4 Building Construction
Rebecca Bach, Evgenia Kanli, Nihat Kiziltoprak

5 Load-Bearing Structure, Fire Protection, Building Physics
Rebecca Bach, Nihat Kiziltoprak, Ulrich Knaack, Marcus Pfeiffer

6 Case Studies
Rebecca Bach, Jerzy Łątka, Fabian Luttropp, Ria Stein

7 Outlook
Robert Götzinger, Ulrich Knaack, Fabian Luttropp, Samuel Schabel

8 Facts and Figures for Engineers
Rebecca Bach, Robert Götzinger, Nihat Kiziltoprak, Marcus Pfeiffer

Glossary
Rebecca Bach, Ulrich Knaack, Samuel Schabel, Ria Stein

Consultation

Prof. Ariel Auslender, TU Darmstadt, Architecture
Prof. Dr. rer. nat. habil. Markus Biesalski, TU Darmstadt, Macromolecular Chemistry and Paper Chemistry
Prof. Dr.-Ing. Andreas Büter, TU Darmstadt, Mechanical Engineering and Plastics Technology
Prof. Dr.-Ing. Dipl. Wirtsch.-Ing. Peter Groche, TU Darmstadt, Production Engineering and Forming Machines
Prof. Dr.-Ing. Ulrich Knaack, TU Darmstadt, Structural Mechanics and Design, Façade Structures
Prof. Dr.-Ing. habil. Stefan Kolling, TU Darmstadt, Structural Mechanics and Design
Prof. Dr.-Ing. Samuel Schabel, TU Darmstadt, Paper Technology and Mechanical Process Engineering
Prof. Dr.-Ing. Jens Schneider, TU Darmstadt, Structural Analysis and Design

Drawings

A major contribution to the success of the book was made by Tessa Krämer,
who created most of the drawings. Other drawings were made by Bianca Biernatek,
Agata Jasiolek and Saphira Wahl.

GLOSSARY

Anisotropic In anisotropic materials, the properties change depending on the direction of the force. One typical example is wood, which can withstand loads differently across the grain than in the direction of the grain.

Area elements are flat components such as plates, diaphragms, shells and membranes. When building with paper, these include ► **Honeycomb boards** and ► **Corrugated multi-wall boards**.

Auxetic is the unusual property of a material that expands when stretched transversely to the stretching direction. Paper is an auxetic material as the paper tapers in the sheet plane when stretched and expands in thickness.

Bending stiffness is the ► **Stiffness** of a slender component related to bending forces. With paper, higher ► **Grammage** can improve the bending stiffness. Sandwich structures such as honeycomb boards have good bending stiffness.

Breaking length means the length that a freely hanging strip of paper must have to tear under its own weight. The value is a commonly used parameter in the paper industry for ► **Tear resistance** since the thickness of a sample or sheet is not required for its determination. The thickness of a sheet of paper is difficult to determine because of the compressibility of paper.

Calenders (from the French *calandre*, "roll") are several rollers arranged one on top of the other, through the gaps of which the paper is passed. The pressure exerted on the paper improves gloss and smoothness while reducing thickness. Calenders are located outside the papermaking machine (offline) or inside the machine between the dryer section and the reel section (online).

CD is the abbreviation for Cross Direction, a description of the axial direction of paper. This refers to the direction in the plane of a sheet perpendicular to the direction of travel on the paper machine ► **MD** (Machine Direction) is offset by 90° to it. Depending on, for example, the fibre orientation, the strength values of the paper differ.

C flute Countless dimensions are possible when designing the flute. Therefore, flutes are classified according to flute type or flute profile, which are defined in DIN 55468-1. They are divided into coarse, medium, fine and microflute, marked with the letters A, C, B and E.

Cobb test After water is applied to paper over a controlled period of time, the water absorption is determined by weighing. This way, the water absorption capacity – i.e. the absorbency of paper materials – can be determined.

Corrugated board consists of three layers (single face corrugated board): thin corrugated paperboard is glued between two layers of paper, and the glue adhesion keeps the corrugation in shape. Corrugated board exploits the principle of the semi-circular arch: the pressure is absorbed by the flutes and distributed evenly to the outer paper board layers. This makes corrugated board much more stable than the three cardboard layers would be without the corrugation.

Corrugated multi-wall board is a sandwich element, the core of which consists of cut-to-size and 90° rotated ► **Corrugated board** glued together in endless webs. Generally, these corrugated cores are covered with ► **Liner paper**. However, due to their intrinsic stability, corrugated cores are also available without liner papers and are used in this form for design purposes.

Corrugated Crush Test (CCT) is a test method for corrugated base papers. The paper is formed into flutes and then loaded by pressure on the face of the paper in a holder that limits bulging or buckling.

Couching is the joining of paper layers under pressure in a wet state, whereby the paper fibres bond via hydrogen bonds. It is a method of producing multi-layer paper or cardboard.

Discretisation is the process of transferring continuous objects into a finite number of discrete objects. It is used to perform both analytical and numerical calculations.

Edge Crush Test (ECT) determines the compressive strength of corrugated board when loaded over two opposite edges. It is measured by pressing a pre-defined section of the board together at its edges, thereby establishing the edge crush resistance. The sample is held between two plates to prevent bulging.

Elongation at break defines a material's possible deformation before breakage. It is the ratio between the material's original length and the initial length after breakage, given in per cent.

Flat crush resistance (FCT – Flat Crush Test) is the ability of a corrugated board to resist a vertically applied force. A sample (100cm²) is placed between plane-parallel compression plates in a crush tester and compressed vertically until the flutes collapse. The greatest resistance that the sample offers in the process gives the value of the flat crush resistance. Fine corrugated board achieves higher values in this process than coarse corrugated board because it has more load-bearing flutes per unit length.

Fluting paper forms the corrugated, middle layer of ► **Corrugated board**. It holds the ► **Liner papers** at a distance. The strength requirements are less stringent for fluting, which is why poorer paper qualities suffice. Fluting is produced based on waste paper.

Fresh fibre materials The most important fresh fibre materials are pulps (without lignin), mechanical wood pulps (with lignin) and semi-chemical pulps. All have advantages and disadvantages. During pulp production, about 50% of the wood mass is "lost" (goes into the solution). The fibres have the best strength values. With mechanic wood pulps, almost no wood is lost; the fibres have lower strengths but higher stiffness and good light diffusion properties. Semi-chemical pulps lie in between.

Grammage is the mass per unit area of a paper, expressed in grams per square metre. Paperboard has a grammage of 250 to 600g/m²; above 600g/m² it is called cardboard.

Honeycomb boards are sandwich-like ► **Area elements**. Contrary to ► **Corrugated boards**, the core of the sandwich element consists of rectangular or hexagonal honeycomb structures, which are covered with liner papers or face sheets.

Isotropic In an isotropic material, its properties are the same in all directions.

Kraft liner is a stable ► **Liner paper** which consists predominantly of ► **Kraft pulp**.

Kraft pulp is a pulp made of high-strength fibres. It is usually produced using the sulphate process, in which cellulose fibres are extracted from the wood of trees by boiling wood chips in a lye solution for several hours. The lignin contained in the wood, as well as polyoses and other wood components, are separated and the fibres remain.

Lignin is a biopolymer, i.e. a macromolecule, which is formed in the cells of perennial plants and makes up 20 to 30% of the cell wall substance of wood. During pulp production, lignin is separated from the wood fibres. Lignin impedes the formation of hydrogen bonds, which are decisive for the strength of papers.

Linear elements are linearly oriented components such as beams or cables, i.e. the length of the element dominates over the width and height of the cross-section. When building with paper, these are typically tubes and cardboard profiles.

Liner paper, also liner or top liner, is the inner or outer covering on one or both sides of corrugated board, between which the wavy corrugated paper is located. A distinction is made between ► **Kraft liner** (made from fresh fibre, predominantly ► **Kraft pulp**), ► **Test liner** (made from recycled paper) and ► **Schrenz** (made from recycled paper of lower strength, mostly used for corrugated paper).

Mass per unit area, also ► **Grammage** specifies the weight of a material related to its dimensions.

MD is the abbreviation for "Machine Direction". Here, the orientation of a sample or a property with respect to ► **CD** (Cross Direction) is offset by 90° in the sheet plane. The strength values of the paper differ depending on the fibre orientation.

Orthotropic The force-deformation behaviour of a material depends on the load direction, i.e. the material has different properties depending on its orientation. Orthotropic is a special form of ► **Anisotropic** behaviour in which the properties are the same in two spatial directions but different in the third.

Partial safety coefficients are required within the framework of stability calculations for structures. For the purpose of higher safety, the stresses are multiplied by partial safety factors. Thus, building material properties that may cause an unfavourable deviation from characteristic values are also taken into account. In the case of paper materials, this allows recording damages that may occur in the construction phase during storage and handling, or that are caused by different manufacturing qualities.

Position stability The position of a component must always be guaranteed, i.e. lifting or rising from the planned position may not occur.

Puncture resistance While ► **Flat crush resistance** determines the deformation of a corrugated board when compressed in the through-thickness direction, puncture resistance determines the ► **Puncturing energy** or the force that is required to push through the ► **Corrugated board** with a standardised penetrator. In the puncture test, coarse corrugated board achieves higher measured values than fine corrugated board as the coarser corrugated board offers greater bending resistance to the pyramid-shaped body during penetration and puncture.

Puncturing energy is the energy used to puncture a corrugated board with a standardised penetrator. See also ► **Puncture resistance**.

Rapid Prototyping is the rapid production of the model of a component according to 3DCAD data (three-dimensional computer-aided design) data. The workpieces are usually produced by additive manufacturing, so-called 3D printing.

Ring Crush Test (RCT) Test method for paper in which a strip is tested for compressive load at the edges. The test strip is placed in a ring-shaped holder to prevent buckling during loading.

Schrenz is a ► **Liner paper** which consists of blended waste paper fibres and has only a low load-bearing capacity.

Semi-chemical pulp is a raw material consisting mainly of mechanically digested wood fibres. In contrast to mechanical wood pulp, the wood chips are pre-cooked in water with digesting chemicals and then mechanically defibrated. The ► **Lignin** is only partially removed in the process, which is why the pulp is not pure. Due to the higher lignin content compared with other pulps, the fibres have higher stiffness and the paper made from them has higher porosities and higher bending stiffness. Semi-chemical pulps are used for the production of paperboard and ► **Fluting paper** used in the production of corrugated board.

Semi-chemical pulp papers consist mainly of wood fibres that have been digested semi-chemically. They must not contain more than 35% recovered paper.

Semi-probabilistic safety concept This safety concept is used in civil engineering for dimensioning in a static calculation. Statistical standard deviations are taken into account both on the action side and on the resistance side, as well as empirical values from realised structures.

Serviceability means that the building is not only structurally sound but also (largely) free from vibrations that would render it unsuitable for use.

Shear resistance is the ability to withstand a tangential load. The shear resistance depends on the type of load and the loaded element or material.

Sizing is a process in papermaking to improve surface resistance and other paper properties. A distinction is made between surface sizing (thin application of sizing to the upper side of the paper web) and internal sizing (addition of ► **Sizing agent** at the wet end before sheet formation).

Sizing agent Historically, so-called animal glues (gelatine solution) or tree resins were used as sizing agents.

Static height denotes the height of the honeycomb or flute cores of ► **Honeycomb boards** and ► **Corrugated multi-wall boards**. Depending on the loading direction, the cores' height and structure are decisive for the structural properties of the ► **Area elements**.

Stiffness This parameter describes the resistance of a body to elastic deformation caused by an external load. The stiffness is determined by the material of the body and its geometry, i.e. shape and size.

Stress-strain diagram The description of a material in terms of its strength, plasticity or brittleness, and elasticity. For this purpose, material samples are tested in a tensile test and the deformation caused by applied stress is determined.

Strip crush resistance (SCT – short compression test) quantifies the resistance of a test strip to compressive stress in the plane direction or compression of the test strip. The clamping length of the specimen in the test apparatus is so short that no kinking or buckling can occur.

Tear resistance is the tensile stress at the moment of tearing of the test specimen. It is determined in tensile tests.

Tensile strength is the load at which a hanging structure breaks, i.e. can no longer transfer loads.

Tensile index, also ► **Breaking length** is the length of a material at which it breaks due to its own weight.

Test liner is a two- or multi-layered ► **Liner paper** made from 100% recovered paper. It is of great importance for corrugated board production. The material is inexpensive, more resilient than ► **Schrenz** but less strong than ► **Kraft liner**.

Torsional buckling is a special case of torsional-flexural buckling in which only twists around the bar axis occur but no lateral displacements. Torsional-flexural buckling is the failure of a beam due to a bending moment, whereby the parts of the supporting structure under compressive stress fail due to buckling.

Volume elements are solid, three-dimensional block-like elements. When building with paper, these include three-dimensional cast shapes such as egg cartons or freely formed components.

ILLUSTRATION CREDITS

COVER Paper Log House, Shigeru Ban, Sherman Contemporary Art Foundation, Sydney, 2017, Photo: Brett Boardman

FRONTISPIECE Aesop DTLA, Brooks + Scarpa, Photo: Art Gray

1 PAPER IN ARCHITECTURE pp. 8–21
Figs. 1, 2 Fabian Luttropp, Tessa Krämer
Fig. 3 Jerzy Łątka
Figs. 4, 5 Jerzy Łątka, Agata Jasiołek
Fig. 6 William Muschenheim/Imageworks, Art, Architecture and Engineering Library, University of Michigan
Figs. 7, 8 Jerzy Łątka, Agata Jasiołek
Fig. 9 Shimizu Yakiu/Shigeru Ban
Figs. 10, 11 Saphira Wahl (after Shigeru Ban)
Figs. 12, 13 Jerzy Łątka
Fig. 14 Elizabeth Felicella/WORKac
Fig. 15 Raymond Adams/WORKac

2 MATERIAL pp. 22–35
Figs. 1–6, 8, 9 Tessa Krämer
Fig. 7 Cynthia Cordt

3 SEMI-FINISHED PRODUCTS AND COMPONENTS pp. 36–53
Figs. 1–7, 14–20 Tessa Krämer
Figs. 8, 9, 12, 13 Tessa Krämer, Sandra Schmidt
Figs. 10, 11 Frederic Kreplin
Fig. 21 Julian Mushövel

4 BUILDING CONSTRUCTION pp. 54–67
Fig. 1 Tessa Krämer, Nihat Kiziltoprak
Figs. 2–6 Tessa Krämer
Figs. 7–15 Tessa Krämer, Evgenia Kanli

5 LOAD-BEARING STRUCTURE, FIRE PROTECTION, BUILDING PHYSICS pp. 68–83
Figs. 1, 3–6, 8, 9, 11, 13–21 Tessa Krämer
Fig. 2 Nihat Kiziltoprak
Fig. 7, 10, 12 Tessa Krämer, Rebecca Bach

6 CASE STUDIES pp. 84–157

Paper House pp. 86–87
Axonometric drawing, floor plan: Tessa Krämer
Photos: Hiroyuki Hirai/Shigeru Ban

Paper Log House pp. 88–89
Axonometric drawing: Tessa Krämer
Elevation: Saphira Wahl, after Shigeru Ban
Photos: Jerzy Łątka

Wikkelhouse pp. 90–95
Exterior drawing, drawings wall build-up and production process: Tessa Krämer
Photos: Yvonne Witte/Fiction Factory
Floor plan, sketch: Oep Schilling

Cardboard School pp. 96–99
Exterior drawing, section: Tessa Krämer
Photos, floor plan, sketch, details: Cottrell & Vermeulen Architecture

Cardboard House pp. 100–101
Axonometric drawing: Tessa Krämer
Photos, sketches: Stutchbury and Pape Architects

Clubhaus Ring Pass Hockey and Tennis Club pp. 102–103
Isometric and detail drawing: Tessa Krämer
Photos: Jerzy Łątka

PH-Z2 pp. 104–107
Axonometric drawing: Bianca Biernatek
Additional drawings, photos: Dratz & Dratz Architekten

Instant Home pp. 108–111
Axonometric drawing: Tessa Krämer
Additional drawings, photos: Leila Dong-Yoon Chu, Fabian Luttropp

Studio Shigeru Ban, KUAD pp. 112–113
Axonometric drawing: Tessa Krämer
Isometric drawing, photos: Jerzy Łątka

House of Cards pp. 114–117
Axonometric drawing: Tessa Krämer
Floor plan: Bianca Biernatek
Photos: Jerzy Łątka

Project team House of Cards: Olga Gumienna, Weronika Lebiadowska, Joanna Malińska, Agata Mintus, Natalia Olszewska, Paulina Urbanik, Damian Wachoński, Magdalena Wiktorska, Wojciech Wiśniewski (Students of Wrocław University of Science and Technology)

TECH 04 pp. 118–119
Axonometric drawing: Tessa Krämer
Floor plan, section: Bianca Biernatek
Photos, detail drawing: Jerzy Łątka
Project team TECH 04: Agata Jasiołek, Agnieszka Gogól, Yana Gvishh, Valeryia Mazurkevich, Martyna Skóra, Martyna Szymańska, Aleksandra Walkowiak (Students of Wrocław University of Science and Technology)

House 01 pp. 120–121
Axonometric drawing: Tessa Krämer
Additional drawings, photos: Evgenia Kanli, Fabian Luttropp

House 02 pp. 122–125
Axonometric drawing: Tessa Krämer
Additional drawings, photos: Fabian Luttropp, Marco Volkmann

Emergency Shelters Made of Paper pp. 126–129
Axonometric drawing: Tessa Krämer
Drawing p. 127, photos: Bianca Biernatek, Alexander Wolf
Drawing p. 128: Saphira Wahl

Cardboard Theatre Apeldoorn pp. 130–131
Axonometric drawing: Tessa Krämer
Additional drawings, photos: Hans Ruijssenaars

Japanese Pavilion, EXPO 2000 pp. 132–133
Axonometric drawing: Tessa Krämer
Exploded isometric: Shigeru Ban
Photo p. 132: Hiroyuki Hirai/Shigeru Ban
Photos p. 133: Jerzy Łątka

Paper Theatre IJburg pp. 134–135
Roof plan: Tessa Krämer
Section, floor plan, elevation: Saphira Wahl
Additional drawings, photos: Mick Eekhout/Octatube

ARCH/BOX pp. 136–137
Drawings: Agata Jasiołek
Photos: Jerzy Łątka

Project team ARCH/BOX: Szymon Ciupiński, Karol Łącki, Dominik Pierzchlewicz, Dominika Jezierska, Martyna Szymańska, Aleksandra Walkowiak, Magdalena Jabłońska, Weronika Lis, Kinga Niewczas, Marcelina Terelak, Aleksandra Wasilenko, Kinga Wasilewska, Julia Pałęga, Jagoda Owsianna, Olga Domalewska, Marcin Dominik, Julia Kochańska, Kacper Kostrzewa, Sara Korżyńska, Przemysław Piorun, Kamil Plich, Michał Sobol, Oliwia Kędzińska (Students of Wrocław University of Science and Technology)

Paper Bridge Pont du Gard pp. 138–139
Isometric drawing: Tessa Krämer
Photos: Theo van Pinksteren/Octatube

PaperBridge pp. 140–141
Photos: Steve Messam

A Bridge Made of Paper pp. 142–143
Isometric drawing: Tessa Krämer
Photos: Evgenia Kanli

Aesop DTLA pp. 144–147
Drawings: Brooks + Scarpa
Photos: Art Gray

Cardboard Bombay pp. 148–151
Photos: Mrigank Sharma/Nudes
Drawing: Nudes

Cardboard Office Pune pp. 152–155
Drawings: studio_VDGA
Photos: Hemant Patil/studio_VDGA

Carta Collection pp. 156–157
Photos: wb form Zürich

7 OUTLOOK pp. 158–169
Figs. 1–3, 5 Fabian Luttropp
Fig. 4 Fabian Luttropp, Marco Volkmann
Figs. 6, 7 Ulrich Knaack
Figs. 8–11 Samuel Schabel

8 FACTS AND FIGURES FOR ENGINEERS pp. 170–187
Fig. 1, 5–13, 15, 18, 19 Tessa Krämer
Figs. 2–4 Tessa Krämer, Marcus Pfeiffer
Fig. 14 Ulrich Knaack (photo); Tessa Krämer (drawing)
Fig. 16 Tessa Krämer (drawing); FRANK-PTI GmbH (photo)
Fig. 17 Tessa Krämer (drawing); Rebecca Bach (photo)
Figs. 20–25 Robert Götzinger, Frederic Kreplin

INDEX